C000091460

Advanced Biomaterials for Orthopaedic Application

Advanced Biomaterials for Orthopaedic Application

The Challenge of New Composites and Alloys Used as Medical Devices

Special Issue Editor

Saverio Affatato

MDPI • Basel • Beijing • Wuhan • Barcelona • Belgrade • Manchester • Tokyo • Cluj • Tianjin

Special Issue Editor
Saverio Affatato
Responsabile Struttura Semplice in Tribologia,
Laboratorio di Tecnologia Medica,
IRCCS Istituto Ortopedico Rizzoli
Italy

Editorial Office
MDPI
St. Alban-Anlage 66
4052 Basel, Switzerland

This is a reprint of articles from the Special Issue published online in the open access journal *Materials* (ISSN 1996-1944) (available at: https://www.mdpi.com/journal/materials/special_issues/ Orthopaedic_Devices).

For citation purposes, cite each article independently as indicated on the article page online and as indicated below:

LastName, A.A.; LastName, B.B.; LastName, C.C. Article Title. *Journal Name* **Year**, *Article Number*, Page Range.

ISBN 978-3-03928-636-2 (Pbk)
ISBN 978-3-03928-637-9 (PDF)

Cover image courtesy of pixabay.com

Contents

About the Special Issue Editor

Saverio Affatato is a Medical Physicist at the IRCCS Istituto Ortopedico Rizzoli (IOR) in Bologna. In this moment, he is the Head of the Wear Characterization of Joint Prostheses Research Unit of the Laboratorio di Tecnologia Medica at IOR (www.ior.it/tecno). He developed new protocols for wear evaluation on hip and knee joint simulator and also particle debris characterisation. He was the scientific and technical lead responsible for some National and European projects (REPO 2000, Eureka 294, Bioker, Nanoker), where the focus of this research was the development of innovative material for orthopaedics applications and their evaluation from a tribological point of view. He has published over 160 peer-reviewed articles in world class high-impact journals that are also well cited (>2200 citations with a corresponding h-index of 25 based on the Scopus platform). He is the referee of some international and prestigious journals such us *Clinical Biomechanics*, *Acta Biomaterialia*, *Biomaterials*, *Proc. IMechE Part H*, and *J. Engineering in Medicine*, etc. He is currently serving as an Editorial Board Member for *J. Health Care Engineering*, *Lubricants*, and *Materials*.

Preface to "Advanced Biomaterials for Orthopaedic Application"

This book is the result of contributions from international researchers active in the various areas of biomaterials. Newly developed biomaterials used as medical devices have led to significant advances not only in regenerative medicine and medical diagnostics but also in in silico fields.

The focus of this book is to provide a review of existing biomaterials in clinical practice, from their properties to recognized issues and further development directions. Many aspects are incorporated in this book, especially in improved medical implants or finite element modelling. Particular emphasis is placed on the simulations of hard-on-soft hip joint prosthesis accounting for dynamic loads calculated from a musculoskeletal model during walking or the development of a novel in silico model to investigate the influence of radial clearance on the acetabular cup contact pressure in hip implants. This book includes considerations of the effects on bone stress, collateral ligament force, and contact stress on other knee compartments of the biomaterials used in this field (biomechanical effects of ultra-high molecular weight polyethylene (UHMWPE) and carbon-fiber-reinforced polyetheretherketone (CFR-PEEK)).

Material responses to different parameters, such as loading patterns, type of motion, and contact velocities, as well as influence of the surrounding environment, which provokes different chemical reactions, can offer valuable further directions for improvement.

This book will provide researchers and professionals with valuable insights into the state-of-the-art developments of biomaterials in clinical practice today from aspects of different areas of expertise: medical, engineering, physics, chemistry, and material science, to motivate them to pursue further research and practical applications.

Saverio Affatato
Special Issue Editor

Article

Effect of Laser Energy Density, Internal Porosity and Heat Treatment on Mechanical Behavior of Biomedical Ti6Al4V Alloy Obtained with DMLS Technology

Żaneta Anna Mierzejewska

Faculty of Mechanical Engineering, Bialystok University of Technology, Wiejska 45c Street, 15-351 Białystok, Poland; a.mierzejewska@pb.edu.pl; Tel.: +48-692-885-870

Received: 13 June 2019; Accepted: 22 July 2019; Published: 22 July 2019

check for updates

Abstract: The purpose of this paper was to determine the influence of selected parameters of Direct Metal Laser Sintering and various heat treatment temperatures on the mechanical properties of Ti6Al4V samples oriented vertically (V, ZX) and horizontally (H, XZ). The performed micro-CT scans of as-build samples revealed that the change in laser energy density significantly influences the change in porosity of the material, which the parameters (130–210 W; 300–1300 mm/s), from 9.31% (130 W, 1300 mm/s) to 0.16% (190 W, 500 mm/s) are given. The mechanical properties, ultimate tensile strength (UTS, Rm) and yield strength (YS, Re) of the DMLS as-build samples, were higher than the ASTM F 1472 standard suggestion (UTS = 1100.13 ± 126.17 MPa, YS = 1065.46 ± 127.91 MPa), and simultaneously, the elongation at break was lower than required for biomedical implants (A = 4.23 ± 1.24%). The low ductility and high UTS were caused by a specific microstructure made of α' martensite and columnar prior β grains. X-Ray Diffraction (XRD) analysis revealed that heat treatment at 850 °C for 2 h caused the change of the microstructure intothe $\alpha + \beta$ combination, affecting the change of strength parameters—a reduction of UTS and YS with the simultaneous increase in elongation (A). Thus, properties similar to those indicated by the standard were obtained (UTS = 908.63 ± 119.49 MPa, YS = 795.9 ± 159.32 MPa, A = 8.72 ± 2.51%), while the porosity remained almost unchanged. Moreover, the heat treatment at 850 °C resulted in the disappearance of anisotropic material properties caused by the layered structure (UTS$_{ZX}$ = 908.36 ± 122.79 MPa, UTS$_{XZ}$ = 908.97 ± 118.198 MPa, YS$_{ZX}$ = 807.83 ± 124.05 MPa, YS$_{XZ}$ = 810.56 ± 124.05 MPa, A$_{ZX}$ = 8.75 ± 2.65%, and A$_{XZ}$ = 8.68 ± 2.41%).

Keywords: Selective Laser Melting; Direct Metal Laser Sintering; porosity; titanium alloys; yield strength; ultimate tensile strength; X-Ray Diffraction

---

1. Introduction

Direct Metal Laser Sintering (DMLS) is predicted to revolutionize the implants manufacturing. However, it is crucial to understand the possibilities and limitations of the process for the Ti6Al4V alloy, which is still the subject of extensive research. Phase transitions, as a result of heat treatment, have a considerable influence on the mechanical properties of the material [1,2], as well as α lath thickness [3]. Hrabe and Quinn [4] conducted a series of experiments to determine the effect of a microstructure on the properties of the samples in order to compare the mechanical properties of the Ti6Al4V alloy. Their work revealed that the orientation of the base plane in relation to the load direction is of great importance. The yield strength was the highest when the direction of the load was perpendicular to the base plane and the lowest in the orientation of an angle of 45°, resulting in a maximum shear stress at the basic plane. They have also proved that the texture has almost no effect on the ductility.

In numerous research papers, it has been proved that the Ti6Al4V alloy's microstructure created with a laser beam has some unique features that are characteristic of DMLS [5]. Tests of Ti6Al4V samples made with the use of Nd: YAG lasers presented that columnar grains are a common feature of microstructures made with additive techniques [6–9]. These grains grow along the boundaries of the original β grains, along with the build direction, as a result of repeated cycles above the β-transus and solidification temperatures, which start at a temperature of about 1660 °C. When passing through the liquidus line, the grain growth of the primary β phase occurs very quickly, making the grains orientation vertical, according to the build direction [10,11]. The laser beam creates a pool of molten material, from which the heat is transferred to the environment and to the already solidified layers. As the material below the top layer is again partially melted, the grains grow epitaxially from the layer below [12–14]. Some researchers suggested that the presence of columnar grains in DMLS samples may result in anisotropic mechanical properties of as-build samples (immediately after sintering) [15,16]. Regarding mechanical properties, it has been shown that cracks tend to spread along columnar grains [17,18]. Depending on the direction of tensile forces and the orientation of the columnar grains, there were differences in the ductility of samples built in a vertical or horizontal direction [7,19,20]. While the tensile strength was approx. 1060 MPa in both directions, the elongation was 11% and 14% for samples build in vertical and horizontal orientation, respectively.

The microstructure of the Ti6Al4V alloy produced by SLM technology is entirely made of the martensitic α'phase [10,21]. This is a result of high rates of heating and cooling of the solidified powder, which are neglected in the DMLS process [22]. Fine martensitic structure α'determines a high yield point and ultimate tensile strength (UTS) of more than 1 GPa, but also has a low elongation value, not exceeding 10%, and therefore lower than forged and casted materials [6,7,10,23]. Vranckenet al. analyzed the effect of various heat treatments on the microstructure of samples manufactured with SLM technology, taking into account different times, temperatures, and cooling rates [24]. The original α'structure has been transformed into a α + β plate mixture with heating below β-transus, while the features of the original microstructure (columnar grains) were preserved. After the heat treatment above the β-transus temperature, a clear change in the microstructure was observed—the columnar β grains were transformed into equiaxed grains.

The presence of defects in the microstructure, which are the source of initiation and propagation of cracks, as well as their random distribution, influences the reduction of fracture strength and fatigue strength due to significant stress peaks [25–28]. The morphology, number, size, and location of defects also affect the durability of samples created with the use of SLM technology. Qui et al., Liu et al., Gong et al., and Zhang et al. proved that spherical defects had less impact on durability if their volume did not exceed 1% [7,26,29,30]. The fracture toughness was significantly lowered, while spherical defects created over 5% of material porosity. Irregular defects in the volume of up to 1% turned out to be much more harmful for the samples tested in terms of mechanical strength. Kasperovich et al. tested samples that were subjected to various types of post-treatment—their aim was crack elimination, which occur in the structure, to improve fatigue strength [31]. Their research indicated that the heat treatment process only affected the change of the microstructure, however, it did not eliminate its defects, and therefore the improvement of fatigue life was negligible. As it is difficult to control the type, quantity, and location of defects in DMLS parts, fatigue strength can be different, even when the manufacturing process is carried out using regular parameters. Therefore, the fatigue strength of components manufactured by SLM technology still raises doubts and requires further research to be improved.

A number of studies conducted by Leuders et al., Qui et al., and Simonelli et al. have shown that it is possible to obtain a density of more than 99.5% for Ti6Al4V samples produced with the use of a laser beam [6,7,10]. Some researchers suggest that, due to the apparent density of the powder (50–60%), small amounts of gas may be entrapped in the material after solidification, resulting in the pores formation, which can be defined as small and empty spaces inside the solid material [3,32]. Gas pores are detrimental to mechanical properties because they act as stress concentrators in components under load, reducing fatigue life and elongation [7,25,33,34]. For this reason, one of the objectives of this work was to analyze the effect of porosity on the mechanical properties of samples obtained with DMLS technology.

The unique properties of titanium in medical applications were discovered relatively recently in the mid-1970s.Pioneering research, which was conducted in 1983, presented that this material promotes bone build-up on its surface (contact osteogenesis), causing a slight inflammatory reaction of tissues to foreign bodies [35]. One of the alloys meeting the requirements for biomaterials is a two-phase alloy Ti6Al4V (Titanium Grade 5/UNS R56400/WNR 3.7165), consisting of aluminum (6%) and vanadium (4%). This alloy is used in many industries, but mainly in biomedical engineering [36].

Concern about the validity of the titanium alloy in biomedical applications arouse the presence of aluminum and vanadium, and its toxicity to living organisms have been well documented [37]. However, specialists in the field of materials and medicine ensure that the occurrence of complications associated with the release of corrosion products to the tissues surrounding the implant is negligible, as titanium and its alloys undergo spontaneous surface passivation [38]. This process is related to the spontaneous formation of a tight and stable TiO_2 layer, formed on the surface, protecting the implant from the aggressive tissue environment and corrosion processes [39]. The unique properties of the Ti6Al4V alloy such as high corrosion resistance, so-called relative strength, low modulus of elasticity, high biocompatibility, excellent osseointegration, or low density, allow for the comprehensive use of its potential [40]. On the basis of ASTM F 1472 specifications for a forged Ti6Al4V alloy used as a medical material, the YS and UTS should not be lower than YS = 868 MPa and UTS = 930 MPa, respectively, and the elongation at the moment of break should be no less than 10%. Obtaining such properties with additive technologies requires a precise selection of DMLS parameters and additional heat treatment, due to which the stresses in the material will be eliminated, and the single-phase martensitic structure will decompose into a two-phase structure suitable for implantable biomaterials.

While the technology of selective laser melting is standardized, each commercial DMLS system presented in the published studies shows a separate characteristic in terms of materials and basic process characteristics (powder delivery to the working field, platform temperature, optical system, and laser) that affect the obtained results [41,42]. Literature analysis indicates a lack of a systematic approach to the study of laser–metal interactions, which would be beneficial for the further development of DMLS. To determine the optimal process parameters, extensive and comprehensive research on the relation of process conditions and product properties are required. Therefore, the key to this study was to investigate how the orientation affects the strength of the material and whether the heat treatment temperature can affect the reduction of the anisotropy of the material obtained with DMLS technology

2. Materials and Methods

The verification of the chemical composition of the analyzed powder was carried out using Thermo ARL Quantris spectrometer Spectrometer (Thermo Fisher Scientific, Waltham, MA, USA). Detailed results of the powder analysis are presented in Table 1. The content of Al and V was within the range specified in ASTMF 2924-14 (Standard Specification for Additive Manufacturing Titanium-6 Aluminum-4 Vanadium with Powder Bed Fusion).

Table 1. Chemical composition and material characteristics of used powder.

Element	Al	V	O	N	H	Fe	C	Ti
wt.%	5.97	4.04	0.195	0.036	0.010	0.24	0.061	Bal.
ASTM F 2924-14	5.5–6.75	3.5–4.5	<0.2	<0.05	<0.015	<0.3	<0.08	Bal.

Powder particle size measurements were performed using the ANALYSETTE 22 Micro Tec plus particle analyzer from FRITSCH GmbH Milling and Sizing (Idar-Oberstein, Germany). According to the powder producer (EOS company, Krailling, Germany), the average particle size is around 30 μm.

However, laboratory analyzes revealed that the particle size was in the range of 10 to 100 μm. The average particle size obtained for three subsequent measurements of the particle distribution was equal to 39.81 μm, with about 94.2% of the particles investigated ranging from 20 to 80 μm (0–30 μm, 18.12%; 30–80 μm, 80.82%, 80–100 μm, and 1.10%). The grains observed under the microscope were spherical and smooth, with numerous satellites, which correspond to the morphology of gas sprayed powder. The advantage of spherical morphology and wide particle size distribution has high fluidity and high packing density.

Thirty groups of test samples were used in experimental research. In each group, different laser powers (130 W, 150 W, 170 W, 190 W, and 210 W) and different beam speeds (300 mm/s, 500 mm/s, 700 mm/s, 900 mm/s, 1100 mm/s, and 1300 mm/s) were used. In all samples, a cross–scan pattern was used and the same distance of scanning vectors was maintained, as well as layer thickness (0.03 mm), spot size (0.1 mm), and hatch distance (0.1 mm). The process parameters used in the present study have been presented in Table 2.

Table 2. Processing parameters.

Scanning Velocity (mm/s)	Laser Power [W]				
	130	150	170	190	210
	Energy Density (J/mm^3)				
300	144	166	188	211	233
500	87	100	113	127	140
700	62	71	81	90	100
900	48	56	63	70	74
1100	40	45	52	58	64
1300	33	38	44	49	54

The energy density (E) was calculated based on laser power (P), the distance between the laser scan line (h), scanning speed (v), and layer thickness (t), which is presented in Equation (1) [13].

$$E = \frac{P}{vht} \; [J/mm^3] \tag{1}$$

DMLS samples were divided into five groups—four of them were heat treated and one remained in the as-build state. Samples were heated for 2 h at 650 °C, 750 °C, 850 °C, and 950 °C. Heat treatment was carried out in a vertical furnace, heating at a rate of about 10 °C/ min, under argon atmosphere (to prevent oxidation of the titanium surface). The samples were cooled with the furnace to 500 °C at a rate of about 0.04 °C/s, and then cooled in the air. After heat treatment, the samples were grinded with abrasive paper with gradations of 200, 500, and 800, purified in 70% ethanol, rinsed with deionized water, and then electropolished in a bath solution using a perchloric acid, glacial acetic acid, and distilled water mixture in a volume ratio of 1:10:1.2 [43]. The current density was equal to 0.3 A/cm^2. The samples were treated with fresh solution at room temperature (25 °C) for 15 min. The Bruker Sky Scan 1172 (Brucker, Billerica, MA, USA) scanner with set parameters: Number of rows—2664, number of columns—4000, Au + CU filter, sample rotation—0.2500, and pixel size—4.28 μm was used for microtomographic optical analysis. Scanned images of samples were reconstructed with the use of software dedicated to micro-CT image analysis—NRecon, Data Viewer, CTvox, and CTAn. Tensile properties were studied using a Hegewald and Peschke INSPEKT (Meß—und Prüftechnik GmbH, Nossen, Germany) test machine with a maximum breaking strength of 5 kN. Displacements were measured by using an extensometer with a 25 mm gauge length. Yield stress and Young's modulus were determined according to ASTM E 111. XRD X–ray phase analysis was performed using a Bruker D8 Advance Eco-ray diffractometer (Brucker, Billerica, MA, USA), equipped with a Cu Kα1 X–ray tube = 1.5406 Å and an SSD160 detector (Brucker, Billerica, MA, USA) adapted for ultra–fast XRD diffraction detection.

3. Results and Discussion

All samples were produced on one mounting plate. Due to the specificity of the process, changes in the dimensional accuracy of the samples with respect to nominal dimensions (Figure 1a) were expected. The dimensional accuracy was affected by the intensive degreasing of samples in an ultrasonic washer and resulted in the rinsing out of loose, unmelted powder grains, and consequently, changes in the dimensions of the samples. Thirty randomly selected samples were subjected to detailed measurements in strictly defined places—fourteen points, marked A through M, presented in Figure 1b.

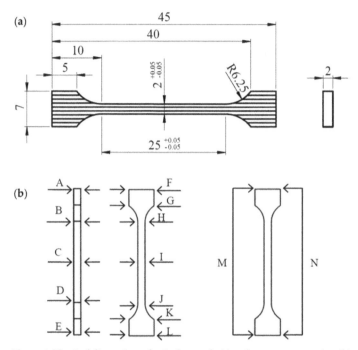

Figure 1. Nominal dimensions of a single sample (**a**) and measurement points (**b**).

The dimensions of each sample were measured with a digital caliper to an accuracy of 0.01 mm. The list of measurements is presented in Table 3.

Table 3. Dimensional accuracy of samples for testing mechanical properties.

Dimension	Thickness A–E [mm]	Necking H/I/J [mm]	Width F/G/K/L [mm]	Length M–N [mm]
Nominal	2	2	7	45
Min	2.48	2.07	6.97	45.55
Max	2.55	2.17	7.16	45.65
Average	2.51	2.12	7.07	45.58
Standard deviation	0.01	0.02	0.06	0.03

3.1. Porosity

Polished and etched surfaces of cylindrical samples were subjected to microscopic observations. The effect can be seen in Figure 2. It was concluded that the change in scanning speed as well as the change in laser power significantly affected the shape and the number of pores.

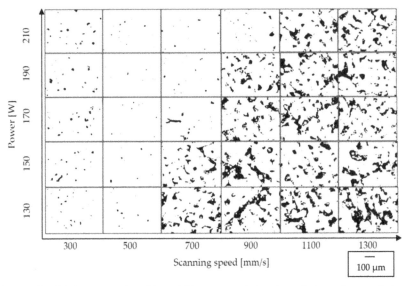

Figure 2. Graphical representation of the variable pore morphology in vertical samples manufactured with Direct Metal Laser Sintering (DMLS) technology, perpendicular to the build direction; 100×.

The quantitative analyzes allowed for a precise assessment of the samples porosity by means of non-destructive microtomographic examination. The correlation between the decreasing value of the melting energy density and the increasing porosity is highly noticeable (Figure 3). However, porosity increases significantly with the increase of scanning speed. This regularity was observed in each group of samples. For samples melted with a laser power of 130 W and a scanning speed of 500 mm/s, the material density was almost 99%. The porosity in the range of 0.72% indicates the presence of internal defects in the structure. An increase is observed below and above 500 mm/s in porosity—from 1.27% at a speed of 300 mm/s to 9.31% at a speed of 1300 mm/s (Figure 3a). Samples melted with a power of 150 W and 300 mm/s were characterized by a porosity of 1.5%. Above 500 mm/s, an increase in porosity was observed—from 3.37% at a speed of 700 mm/s to 8.43% at a speed of 1300 mm/s. The minimum value porosity of 0.46% was observed at a laser speed of 500 mm/s. In comparison to samples manufactured at 130 W with the same speed, the porosity was lower by 0.26%. The difference between the maximum porosity values obtained for the highest laser speed was less than 1% (Figure 3b). For samples melted with a laser power of 170 W at scanning speeds from 300 to 700 mm/s, the porosity of the material varies from 1.69% to 0.8%. Above 700 mm/s, an increase in porosity is observed—from 5.29% at a speed of 900 mm/s to 7.29% at a speed of 1300 mm/s. Samples melted with a power of 170 W achieved a density of more than 99% for two scanning speeds—500 and 700 mm/s. The highest density, 99.73%, was obtained for samples scanned with 700 mm/s (Figure 3c). Additionally, for samples melted with a laser power of 190 W and scanning speed of 500 mm/s and 700 mm/s, porosity did not exceed 1%. The lowest value of porosity was observed for samples melted with 500 mm/s—0.16%. At a scanning speed of 300 mm/s, microstructure defects in the analyzed samples were slightly below 2%, while a further increase in scanning speed above 500 mm/s resulted in higher porosity—from 0.68% for 700 mm/s to 6.57% for 1300 mm/s (Figure 3d). Similar characteristics were found for samples melted with a power of 210 W. The porosity obtained at 300 mm/s exceeded 2% and a decreasing tendency was observed along with the increase of sintering speed to 700 mm/s. After exceeding 700 mm/s, the porosity increased again, reaching a maximum value of 6.16% at a sintering speed of 1300 mm/s. The lowest porosity was obtained at 700 mm/s and was 0.45% (Figure 3e).

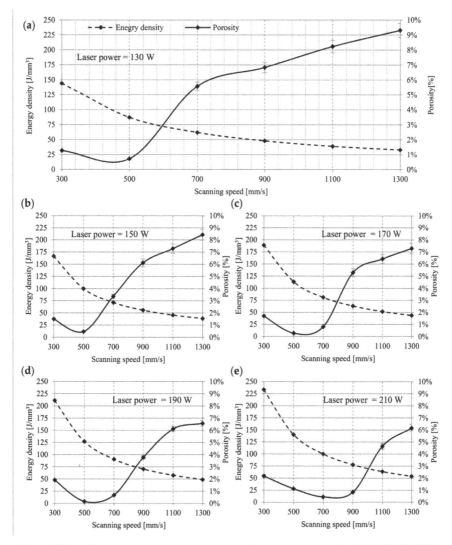

Figure 3. The influence of scanning speed and energy density on the porosity: (**a**) 130 W, (**b**) 150 W, (**c**) 170 W, (**d**) 190 W, (**e**) 210 W.

At low energy density (33–71 J/mm^3), the porosity ranges from 9.31% to 3.37%. The pores are unevenly distributed, irregular in shape, and interconnected. Porosity is characterized by large recesses filled with loose particles of unmelted grains. A possible explanation for this is a low energy density and relatively small depth of laser penetration, which make the size of the melt pool too small as the powder particles are not sufficiently liquefied to completely melt, but provides a sufficient bond between the layers. The increase in energy density from 78 J/mm^3 to 127 J/mm^3 generates a relatively high temperature, which facilitates the flow of liquid and fills the space between already melted grains, causing the porosity in the range of 0.84–0.16%.

At higher energy densities, the pores are small and mostly spherical, and their formation is most often associated with gas bubbles trapped under the molten powder layer. An increase in the laser energy density above 127 J/mm^3 causes changes in the pore morphology and an increase in porosity from 1.12% for 140 J/mm^3 to 2.18% for 233 J/mm^3 (Figure 4).

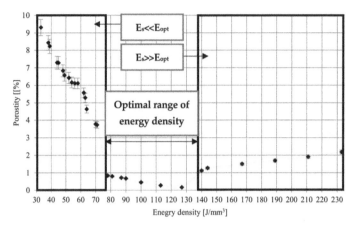

Figure 4. Relation of porosity and energy density for Ti6Al4V DMLS alloy.

The influence of energy density on porosity was also investigated by Han et al. [44], Laquai et al. [45], and Dilip et al. [46]. While the range of the optimal power density indicated by Han and Lagui is similar to that obtained in the present study (120–180 J/mm^3 and 120–195 J/mm^3, respectively), significant differences were observed. Both authors state that the porosity in the tested samples did not exceed 0.05%. What is more, Han also revealed that, in the range of much lower (60 J/mm^3) and much higher (240 J/mm^3) energy density, the density of melted material is not less than 99.75%. Lagui, on the other hand, observed an increase in porosity just above 1% for only energy densities below 50 J/mm^3 and above 300 J/mm^3. Completely different characteristics were presented by Dilip et al. [47]. They showed that the optimal range of energy density was between 50 and 66 J/mm^3 (porosity below 0.5%), for energy densities around 40 J/mm^3, the porosity increased to about 5%, while at higher energy densities between 90 and 130 J/mm^3, the porosity was about 10%.

3.2. Mechanical Properties

The layered structure obtained in DMLS technology makes the samples characterized by anisotropic mechanical properties. For this purpose, two types of samples were printed—vertically oriented samples (V, ZX) and horizontally orientated samples (H, XZ; Figure 5). The layering in both samples was perpendicular to the building direction (Z), while the tensile forces acting on the samples were oriented perpendicular to the layers in the ZX samples and parallel to the layers in the XZ samples. The tensile strength results presented in Figures 6–10 were calculated on the basis of five consecutive tests.

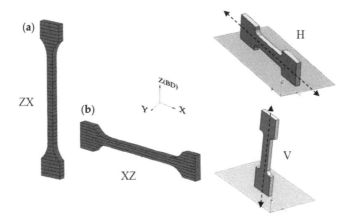

Figure 5. Tensile samples oriented in: (a) XZ plane and (b) ZX plane.

3.2.1. As-build Samples

Based on ASTM F 1472 specifications for the forged Ti6Al4V alloy used as medical materials, YS and UTS should not be lower than 868 and 930 MPa, respectively, and elongation at break should be no less than 10%. For all as-built samples (Figure 6a–c), the yield strength and tensile strength significantly exceeded the minimum requirements for the forged material described in the standard and were equal to 1065.46 ± 127.91 MPa and 1100.13 ± 126.17 MPa, respectively. The highest strength was characterized by samples melted with 190 W power and 500 mm/s speed (Re max = 1261 ± 5.65 MPa and Rm max = 1288 ± 28.28 MPa), while the lowest was the samples melted at 130 W and a speed of 1300 mm/s (Re min = 885 ± 7.07 MPa; Rm min = 918.5 ± 28.99 MPa). In the group of samples melted with 130 W and 1100–1300 mm/s, strength parameters did not meet the criteria of ASTM F 472.The fragility of the martensitic microstructure and the presence of residual stresses are responsible for the lower ductility.

It can be determined that the samples elongation was definitely lower than required in this standard, and its average value was 4.23 ± 1.24%, with the maximum elongation in this group of samples being A max = 5.8 ± 0.84%, and the minimum A min = 2.2 ± 0.56%. The tensile strength of Ti6Al4V samples produced by the DMLS method is only slightly higher than their yield strength, which indicates the brittleness of the material and its low ductility. An increase scanning speed above 700 mm/s in almost every sample group influences the reduction of the strength parameters and elongation. This is due to the fact that higher speeds cause the generation of a lower energy density, which results in structural defects and the weakening of bonds between layers. Additionally, the increase in power affected the changes of mechanical parameters—tensile strength increased from 1050 ± 136.97 MPa (130 W) to 1148.83 ± 106.74 MPa (210 W). YS and A also increased—from 1013.58 ± 134.95 MPa to 1112.41 ± 109.85 MPa and from 3.72 ± 1.43% to 4.72 ± 0.99%.

The surfaces fracture of non-heat treated samples showed mixed characteristic of brittle and ductile fracture, accompanied by very small plastic deformation. Fracture surfaces were characterized by the presence of shallow dimples on quasi cleavage surface, delamination, or stepped cracks, suggesting intergranular destruction along hard, prone to cracking, brittle needles α', which then propagated due to decohesion caused by further deformation of the sample. A large number of gas defects and pores caused the concentration of stress and violent crack propagation, and the mixed nature of cracking to a limited elongation (Figure 6d).

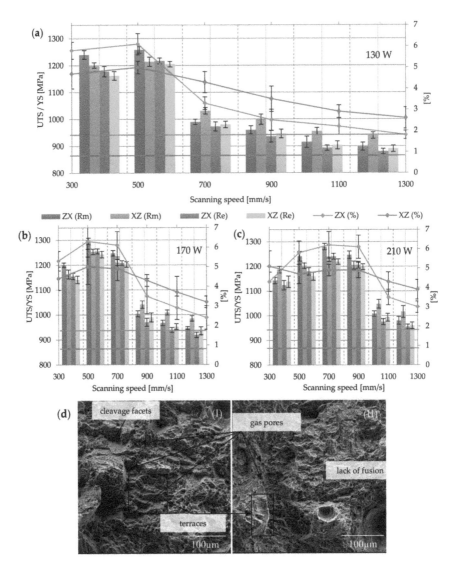

Figure 6. Strength parameters of as–built samples with different laser speeds and powers: (**a**) 130 W, (**b**) 170 W, (**c**) 210 W, and (**d**) fracture surface: (I) 130 W, 300 mm/s, ZX; (II) 130 W, 300 mm/s, XZ; 900×; Rm min — and Re min — for Ti6Al4V in accordance with ASTM F 1472.

3.2.2. Heat Treatment at 650 °C

Despite heat treatment, the results of tensile and yield strength were still higher than the minimum determined by the ASTM F 1472 standard (Rm = 1071.7 ± 128.08 MPa; Re = 994.85 ± 124.91 MPa), and the elongation was also unsatisfactory and almost a half lower than required (A = 4.24 ± 1.25%). The highest strength was characterized by samples melted with the power of 190 W and the speed of 500 mm/s (Rm max = 1237 ± 26.87 MPa; Re max = 1192.5 ± 6.36 MPa), the lowest was the samples melted with the power of 130 W at a speed of 1300 mm/s (Rm min = 864.5 ± 24.74 MPa; Re min = 810 MPa). The tensile strength of ZX samples decreased to an average of 1050.73 ± 145.97 MPa, and XZ samples to 1023.86 ± 206.64 MPa. With the decrease in strength properties, the elongation increased for ZX samples to 4.27 ± 1.57%, and for XZ samples to 4.21 ± 0.83%. The differences in tensile strength between the samples printed in the XZ plane and ZX plane was approx. 3.3% (Figure 7a–c). Obtained results clearly indicate that annealed samples had higher elongation at break—regardless of their orientation, and thus better ductility, in comparison to as-build samples. A large range of values from average elongation at break indicates that the improvement is not as significant as in other works presented in the literature [24,25,45]. The minimum elongation value, according to the standard, cannot be lower than 10%, which in the case of the applied heat treatment, was not obtained for any of the samples, therefore reducing the stresses turned out to be beneficial for DMLS Ti6Al4V, but is still insufficient to meet the criteria imposed by ASTM F 1472. The surface of the breakthrough remained mixed, with features of brittle and ductile fractures, with shallow dimples staircase cracks and pores (Figure 7d).

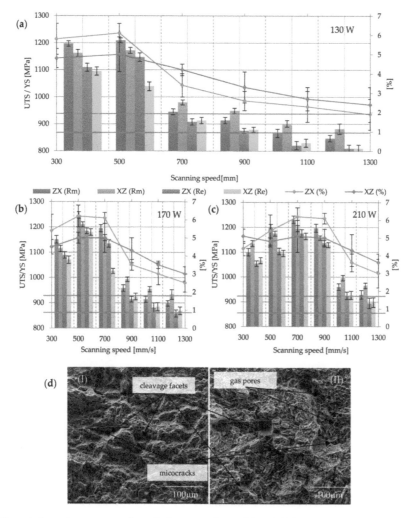

Figure 7. Strength parameters of samples heat treated at 650 °C with different laser speeds and powers: (a) 130 W, (b) 170 W, (c) 210 W, and (d) fracture surface: (I) 130 W, 300 mm/s, ZX; (II) 130 W, 300 mm/s, XZ; 900×; Rm min — and Re min — for Ti6Al4V in accordance with ASTM F 1472.

3.2.3. Heat Treatment at 750 °C

Annealing at 750 °C for 2 h significantly reduced the strength parameters and increased the plasticity of the material, as a result of partial phase transformation $\alpha' \rightarrow \alpha + \beta$ and at the same time caused the growth of grain α, thus changing the phase composition and grain morphology (in relation to as-build samples).

For heat treatment at 750 °C, the Rm of the samples was equal to 1002.88 ± 119.42 MPa, while Re = 900.07 ± 77.43 MPa. The highest strength was again achieved in samples melted with a power of 190 W and a speed of 500 mm/s (Rm max = 1169.5 ± 6.36 MPa; Re max = 1085.5 ± 13.43 MPa). The lowest tensile strength and yield strength were obtained in samples melted with a power of 130 W and a speed of 1300 mm/s again—the average value of Rm min = 812 ± 11.31 MPa; Re min = 723 ± 12.72 MPa. Heat treatment also had a slight influence on the elongation—increased from A = 5.24 ± 1.25% to A = 5.71 ± 1.69%. The difference in the strength of the ZX and XZ samples also decreased: ZX Rm = 1002.27 ± 126.06 MPa, XZ Rm = 998.72 ± 113.82 MPa, similar to the yield strength—Re = 899.53 ± 133.26 MPa (for ZX), Re = 900.6 ± 115.76 MPa (for XZ). The elongation for ZX samples increased to 5.75 ± 2.07%, while for XZ samples, to 5.67 ± 1.23% (Figure 8a–c). The nature of the fractures area has changed slightly, showing more features of a ductile fracture than in the case of previous samples. On the fracture surfaces, defects of the microstructure, i.e., pores, cracks, and discontinuities of the material, were visible (Figure 8d).

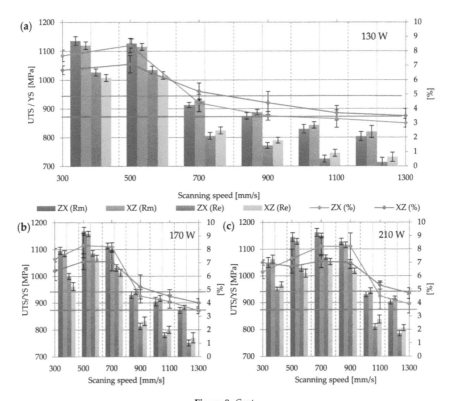

Figure 8. *Cont.*

Figure 8. Strength parameters of samples heat treated at 750 °C with different laser speeds and powers: (**a**) 130 W, (**b**) 170 W, (**c**) 210 W, and (**d**) fracture surface: (I) 130 W, 300 mm/s, ZX; (II) 130 W, 300 mm/s, XZ; 900×; Rm min — and Re min — for Ti6Al4V in accordance with ASTM F 1472.

3.2.4. Heat Treatment at 850 °C

Heat treatment at 850 °C led to a change in the microstructure and decomposition of the α′phase to the α + β. Therefore, it can be assumed that the samples annealed at this temperature presented the best relationship between strength and elongation. The average tensile strength value for all analyzed samples was equal to 908.63 ± 119.49 MPa, yield strength 795.9 ± 159.32 MPa, and elongation at break 8.72 ± 2.51%. The highest tensile and yield strength were again characterized by samples of 190 W/500 mm/s (Rm max = 1058 ± 4.94 MPa; Re max = 958 ± 4.95 MPa), the lowest was again the samples of 130 W/1300 mm/s. The average value of Rm min was 724.5 ± 4.95 MPa, and Re min = 625.5 ± 6.36 MPa. A significant part of the samples treated at this temperature does not meet the criteria of ASTM F 1472 in terms of tensile strength and yield strength, but they still meet the criteria of ASTM F 1108-14.

What is more, processing at 850 °C resulted in homogenization of the microstructure, and as a result, anisotropic properties of differently orientated samples have disappeared. The tensile strength of ZX samples was therefore Rm = 908.36 ± 122.79 MPa, samples XZ Rm = 908.97 ± 118.198 MPa, the yield strength Re for both orientations was equal to: 807.83 ± 124.05 MPa and 810.56 ± 124.05 MPa, and elongation A: 8.75 ± 2.65% and 8.68 ± 2.41% (Figure 9a–c). The surface fracture is typically ductile, with dimples and voids, in the middle of which there are particles of inclusions. In the central part of the crack, the dimples are equi-axial, while on the sides they have a strong shearing character and a parabolic shape (Figure 9d).

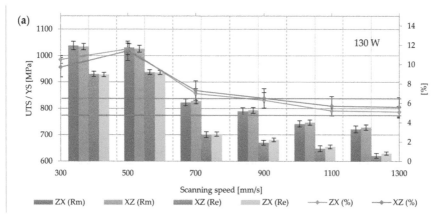

Figure 9. *Cont.*

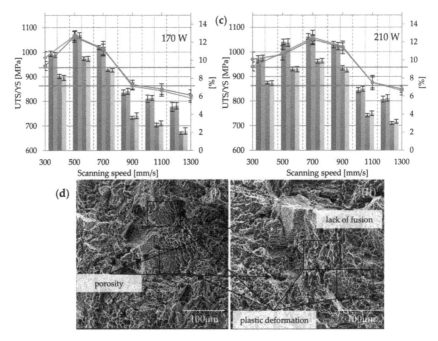

Figure 9. Strength parameters of samples heat treated at 850 °C with different laser speeds and powers: (a) 130 W, (b) 170 W, (c) 210 W, and (d) fracture surface: (I) 130 W, 300 mm/s, ZX; (II) 130 W, 300 mm/s, XZ; 900×; Rm min —— and Re min —— for Ti6Al4V in accordance with ASTM F 1472.

3.2.5. Heat Treatment at 950 °C

Heat treatment at 950 °C significantly affected the change in the microstructure—similarly to that in the lower temperature treatment, the α′martensitic phase decomposed into the α + β phase mixture, but temperature close to β-trasus caused increased grain size and deterioration of plastic properties. Therefore, along with the decrease in the value of strength properties, the elongation at break decreases. As a result, the analyzed properties have the following values: Tensile strength Rm for all analyzed samples is 848.48 ± 86.56 MPa, yield strength 781.35 ± 96.3 MPa, and elongation at break 7.75 ± 2.69%. The highest measured values of tensile strength and yield stress are, respectively, Rm max = 931 ± 1.41 MPa; Re max = 927.5 ± 3.54 MPa, while the lowest—Rm min = 702.5 ± 12.5 MPa and Re min = 615.5 ± 6.36 MPa (Figure 10a–c). The character of the fracture surface remains ductile (Figure 10d).

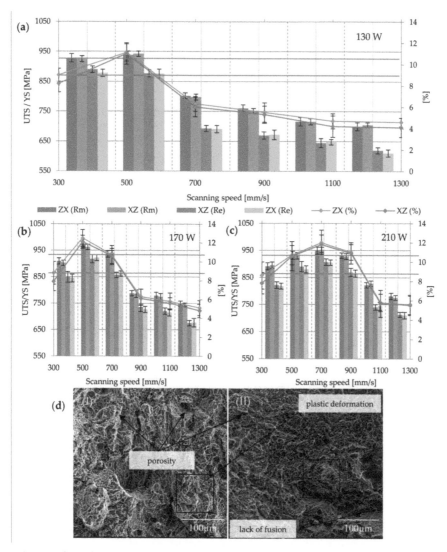

Figure 10. Strength parameters of samples heat treated at 650 °C with different laser speeds and powers: (a) 130W, (b) 170W, (c) 210W, and (d) fracture surface: (I) 130 W, 300 mm/s, ZX; (II) 130 W, 300 mm/s, XZ; 900×; Rm min — and Re min — for Ti6Al4V in accordance with ASTM F 1472.

The obtained data is supplemented by the assessment of the influence of porosity on the strength and elongation of samples, the yield point and Young's modulus (Figures 11 and 12). With the increase of porosity, all these parameters revealed a downward trend, which clearly indicates that defects of the microstructure strongly affect the mechanical properties of the material.

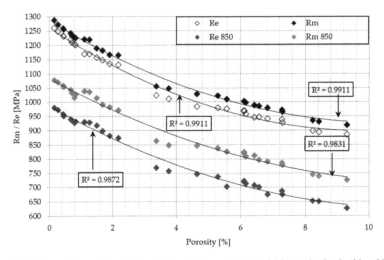

Figure 11. Influence of porosity on Ultimate Tensile Strength and Yield Strength of as-build and heat treated at 850 °C samples.

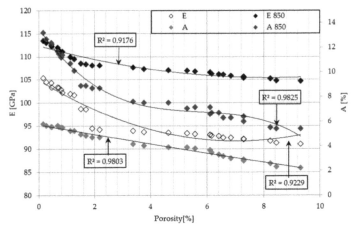

Figure 12. Influence of porosity on Young modulus and elongation of as-build and heat treated at 850 °C samples.

The analysis of the obtained results revealed that samples build in the ZX orientation compared with XZ had a higher yield strength, tensile strength, and elongation, if the volume of the microstructure defects does not exceed 2%. Vertical samples (ZX)—due to the layered structure—are seemingly more prone to initiation and cracks propagation than horizontal samples (XZ), however, they showed higher tensile strength and yield strength. The explanation of this phenomenon is related to the microstructure, especially to the columnar β grains. These grains grown perpendicular to successive layers of molten powder and in accordance with the direction of building the element. As a result, in the XZ samples, tensile force direction was perpendicular to the β grains, whereas in the ZX samples, the direction of the force was parallel. In vertical samples, the grains were axially extended, which had a positive effect on the strength properties of the whole sample. The same orientation between the columnar grains and the direction of the tensile forces led to a reduction of the grain boundary resistance, which no longer constituted an effective obstacle to the dislocation movement and which could be subject to greater distortion [48,49]. Reducing the energy density by reducing the laser beam power and increasing the scanning speed resulted in defects between successive layers of material [50,51]. For highly porous samples, force was applied parallel to the layers and caused pores deformation consistent with the direction of force, whereby the stress concentration around the pores was several times lower, and consequently, the material was less susceptible to the initiation and propagation of cracks.

Based on the microstructure and changes in mechanical properties analysis, presented in the paper, DMLS parameters, application of which guarantees the achievement of a biomaterial with properties consistent with ASTM F 1142, were indicated in Table 4.

Table 4. List of DMLS parameters and properties of Ti6Al4V alloy heat treated at 850 °C meeting the criteria of ASTM F 1142 standard for implantable biomaterials.

P [W]	V [mm/s]	Rm [MPa]		Re [MPa]		A [%]		E [GPa]		Es [J/mm³]	P [%]
		ZX	XZ	ZX	XZ	ZX	XZ	ZX	XZ		
150	500	1056 ± 13	1052 ± 12	954 ± 10	950 ± 8	13.4 ± 0.5	13.2 ± 0.6	106.5 ± 1.3	106.2 ± 0.6	100	0.46
170	500	1074 ± 13	1069 ±10	974 ± 9	973 ± 7	13.7 ± 0.6	13.6 ± 0.5	107.5 ± 0.7	107.5 ± 0.8	113	0.27
190	500	1080 ± 10	1077 ± 11	982 ± 9	980 ± 9	14.3 ± 06	14.1 ± 0.7	108.0 ± 1.2	108.7 ± 0.5	127	0.16
210	700	1062 ± 12	1055 ± 10	962 ± 9	954 ± 8	13.5 ± 0.5	13.3 ± 0.8	107.2 ± 1.0	107.4 ± 0.9	100	0.45

3.3. Microstructure Characterisation

Unique thermal features, such as heat input, large temperature gradients, and high cooling ratesdramatically affected as-built martensitic α′microstructures and led to high residual stresses. Annealing at 650 °C for 2 h was intended to remove residual stress from the material. The use of this heat treatment resulted in the decomposition of a small amount of the martensitic phase in the α + β equilibrium phase, as well as the nucleation and slow increase of the α phase at the boundaries of martensitic needles. Nevertheless, the microstructure has not changed significantly compared to untreated, and the columnar grain structure has been retained. This is due to the fact that rapid cooling during the DMLS process leads to the formation of dislocation within the alloy. These defects have an inhibitory effect on the α-phase distribution. Furthermore, during annealing at 650 °C, which is significantly below than β–transus, the driving force of the phase distribution is insufficient. The microstructure after heat treatment at 650 °C still remained needle-shaped (Figure 13a). The only noticeable change was the increase in grain thickness to 0.6 μm (Table 5).

The microstructure of samples annealed at 750 °C for 2 h and cooled together with the furnace clearly indicate the decomposition of fine martensite into α and β mixture, in which the α phase occurs in the form of fine needles (Figure 13b). The analyzed microstructure consisted of a smaller amount of α′ martensite embedded in the more stable α and β phase. According to the findings of Shunmugavel et al., the upper limit of martensite decay in the Ti6Al4V alloy is 800 °C, which means that only above this temperature martensite is completely decomposed into the phases α and β [14]. The average thickness of α′ plates was 1.2 μm.

Heating of samples at 850 °C for 2 h significantly influenced the changes of the microstructure—the α′ phase completely decomposed into the mixture of α and β phases. The α grain thickness increased to the average value 1.5 μm. Heat treatment carried out at 850 °C did not change the morphology of the primary β grains. The reason for this is the coexistence of the α + β phases, which mutually inhibit their further growth. Different orientation of the growth of adjoining grains prevents the migration of boundaries in the axial direction, which leads to slight changes in their shape (Figure 13c).

The increase of the annealing temperature to 950 °C resulted in the reduction of the inhibitory effect of the grain boundaries—the structure was transformed into coarse grains. The average width of the plate grains of the α-phase after annealing was approximately 2.3 μm. This was the result of the increase in the driving force of the α → β phase decay and the joining of the adjacent β grains under the influence of high temperature. In the fully lamellar microstructure, the size of the α colony and the alternately packed α and β plates with a clear orientation define the grain and have a strong influence on the size and length of the slip, determining the ductility of the Ti6Al4V alloy (Figure 13d). Thus, after a heat treatment at 950 °C, a lower strain was expected during the static tensile test.

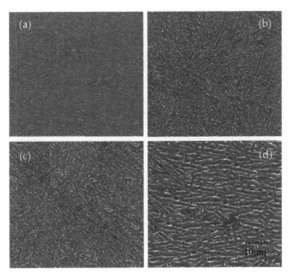

Figure 13. SEM images of microstructure of heat treated samples: (**a**) 650 °C, (**b**) 750 °C, (**c**) 850 °C, and (**d**) 950 °C.

Table 5. Change the grain size in respect of energy density and temperature.

Heat Treatment	Energy Density [J/mm³]			
	33	100	233	Average
AB	0.17 ± 0.05	0.25 ± 0.08	0.38 ± 0.06	0.27 ± 0.11
650 °C	0.42 ± 0.11	0.57 ± 0.12	0.81 ± 0.11	0.60 ± 0.22
750 °C	0.81 ± 0.10	1.16 ± 0.15	1.63 ± 0.09	1.20 ± 0.41
850 °C	1.17 ± 0.12	1.49 ± 0.13	2.11 ± 0.11	1.59 ± 0.48
950 °C	1.83 ± 0.15	2.31 ± 0.18	2.80 ± 0.21	2.31 ± 0.49

The analysis of the microstructure made of Ti6Al4V alloys produced under heat treatment was conducted by phase analysis (Figure 14). XRD spectra clearly presented that the α′ phase was the dominant fraction in as-build samples. All peaks untreated material were indexed as the α′ phase, while the β phase was not detected. Both martensitic phases α and α′ have similar crystallographic cell parameters. However, since fast cooling may promote martensitic transformation, it was expected that the hexagonal hcp phase in the material was acicular martensite. Therefore, metastable martensite remained as the only phase present at room temperature in the alloy Ti6Al4V produced by DMLS technology. In the material heat treated at 650 °C, the main peaks remained the same as in the as-build, however peaks (110) β-Ti2θ = 38.57° and (211) β-Ti (2θ = 71.02°) corresponded to the β phase. A similar pattern was obtained for heat treatment at 750 °C, only the intensity of α–peaks increased due to the increase in grain thickness. Figure 9 presented phase analysis for V oriented samples melted with an energy density of 100 J/mm³.

Heat treatment at 850 °C resulted in complete decomposition of α′ martensite into α + β phase and this influenced the XRD pattern obtained in the study, because peaks 2θ = 35.40°, 39.37°, 40.47°, 53.31°, and 63.54° corresponded to the plate α-phase, not the α′ phase. The peaks of this phase showed strong intensity, which was related to the increase in grain size. The intensity of peaks (110) β-Ti

(2θ = 38.57°) and (211) β-Ti (2θ = 71.02°) also increased, which clearly indicates that a significant amount of β phase increased due to the redistribution of alloying elements. The XRD of the annealed material at 950 °C showed that the microstructure of the DMLS alloy formed a mixture of α + β phases. This unambiguously confirms that the transformation temperature of β–transus is higher than 950 °C.

The change of microstructure as a result of heat treatment affects not only mechanical properties, but also corrosion resistance. According to the literature, as-build samples show a poorer corrosion resistance compared to heat treated samples. Based on published studies, it can be concluded that the corrosion resistance depends on the volume of β phase—the more that is in the structure, the higher the corrosion resistance [52,53].

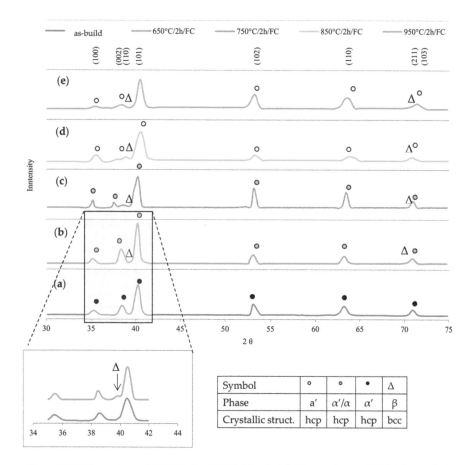

Figure 14. XRD spectrum of Ti6Al4V alloy DMLS: (**a**) as-build, (**b**) annealing at 650 °C, (**c**) annealed at 750 °C, (**d**) 850 °C, and (**e**) 950 °C for 2 h, cooled with a furnace.

4. Conclusions

This work aimed to determine the influence of DMLS parameters on the resulting properties of the obtained material and to assess the effect of different heat treatment temperatures on the structure and mechanical properties of samples oriented vertically and horizontally. The results of completed research are the following conclusions:

- The results of porosity in samples made with DMLS technology differs from those presented in the literature; however, it should be noted that the micro-CT porosity analysis, which was carried out in this work, allowed the assessment of total porosity, whereas in studies presented in the data of the reference, these measurements were made on the basis of fragmentary image analysis or the Archimedes method, which allows the estimation of apparent density.

- At low energy density (33–71 J/mm^3), the porosity varies from 9.31% to 3.37%; the increase in energy density from 78 J/mm^3 to 127 J/mm^3 causing the porosity in the range of 0.84–0.16%; above 127 J/mm^3, the porosity increase again as the effect of the material overheating.

- The microstructure created in the DMLS process is fine-grained α' martensite, which determines high strength properties and low ductility; this microstructure—due to the ferromagnetic properties of martensite—should not be considered as a biomaterial and requires operations whose aim is to change the microstructure to a biphasic phase and to obtain parameters compliant with the normative ones.

- The tests showed that the most favorable combination of mechanical properties and structure without the α' phase may be obtained by annealing the material in the temperature range 850–950 °C for 2 h, cooled together with the furnace.

- After heat treatment, the samples showed YS and UTS similar to those for wrought and annealed Ti6Al4V—this was the effect of the coexistence of the α and β phases.

- Tensile strength tests showed the sensitivity of the as-build material for porosity and orientation of samples, which results from the creation of columnar grains growing along the boundaries of the prior β grains.

- Heat treatment influenced the reduction of tensile (Rm) and yield strength (Re) parameters with a simultaneous increase of elongation (A) and Young's modulus (E), additionally, heat treatment at 850 °C resulted in homogenization of the microstructure and elimination of anisotropy resulting from different directions of stretching samples and layering.

- The highest density of samples (porosity not exceeding 0.5%) was obtained for samples melted with energy density in the range of 100–127 J/mm^3.

- The analysis of the obtained results revealed that samples build in the ZX orientation compared with XZ had a higher yield strength, tensile strength, and elongation, if the volume of microstructure defects does not exceed 2%.

- Regardless of the energy density, the microstructure obtained in the DMLS process consists of martensite needles, however, the higher the energy density, the larger the grain size; the size of grains also grows with the increasing temperature of heat treatment, which affects the reduction of strength properties and increase of elongation.

Funding: This research received no external funding.

Acknowledgments: Studies have been financed from the funds for science of Ministry of Science and Higher Education of Republic of Poland, project no. No S/WM/1/2017.

Conflicts of Interest: The authors declare no conflict of interest.

References

1. Wu, S.Q.; Lu, Y.J.; Gan, Y.L.; Huang, T.T.; Zhao, C.Q.; Lin, J.J.; Lin, J.X. Microstructural evolution and microhardness of a selective-laser-melted Ti-6Al-4V alloy after post heat treatments. *J. Alloy. Compd.* **2016**, *672*, 643–652. [CrossRef]

2. Popovich, A.; Sufiiarov, V.; Borisov, E.; Polozov, I. Microstructure and mechanical properties of Ti-6Al-4V manufactured by SLM. *Key Eng. Mater.* **2015**, *651*, 677–682. [CrossRef]

3. Galarraga, H.; Lados, D.A.; Dehoff, R.R.; Kirka, M.M.; Nandwana, P. Effects of the microstructure and porosity on properties of Ti-6Al-4V ELI alloy fabricated by electron beam melting (EBM). *Add. Manuf.* **2016**, *10*, 47–57. [CrossRef]

4. Hrabe, N.; Quinn, T. Effects of processing on microstructure and mechanical properties of a titanium alloy (Ti-6Al-4V) fabricated using electron beam melting (EBM), Part 2: Energy input, orientation, and location. *Mater. Sci. Eng. A* **2013**, *573*, 271–277. [CrossRef]

5. Sallica-Leva, E.; Caram, R.; Jardini, A.L.; Fogagnolo, J.B. Ductility improvement due to martensite α′ decomposition in porous Ti-6Al-4V parts produced by selective laser melting for orthopedic implants. *J. Mech. Behav. Biomed.* **2016**, *54*, 149–158. [CrossRef] [PubMed]

6. Leuders, S.; Vollmer, M.; Brenne, F.; Tröster, T.; Niendorf, T. Fatigue strength prediction for titanium alloy TiAl6V4 manufactured by selective laser melting. *Metall. Mater. Trans. A* **2015**, *46*, 3816–3823. [CrossRef]

7. Qiu, C.; Adkins, N.J.; Attallah, M.M. Microstructure and tensile properties of selectively laser-melted and of HIPed laser-melted Ti-6Al-4V. *Mater. Sci. Eng. A* **2013**, *578*, 230–239. [CrossRef]

8. Vandenbroucke, B.; Kruth, J.P. Selective laser melting of biocompatible metals for rapid manufacturing of medical parts. *Rapid Prototyping J.* **2007**, *13*, 196–203. [CrossRef]

9. Calignano, F. Design optimization of supports for overhanging structures in aluminum and titanium alloys by selective laser melting. *Mater. Des.* **2014**, *64*, 203–213. [CrossRef]

10. Simonelli, M.; Tse, Y.Y.; Tuck, C. Effect of the build orientation on the mechanical properties and fracture modes of DMLS Ti-6Al-4V. *Mater. Sci. Eng. A* **2014**, *616*, 1–11. [CrossRef]

11. Xu, W.; Sun, S.; Elambasseril, J.; Liu, Q.; Brandt, M.; Qian, M. Ti-6Al-4V additively manufactured by selective laser melting with superior mechanical properties. *JOM* **2015**, *67*, 668–673. [CrossRef]

12. Kok, Y.; Tan, X.P.; Wang, P.; Nai, M.L.S.; Loh, N.H.; Liu, E.; Tor, S.B. Anisotropy and heterogeneity of microstructure and mechanical properties in metal additive manufacturing: A critical review. *Mater. Des.* **2018**, *139*, 565–586. [CrossRef]

13. Sallica-Leva, E.; Jardini, A.L.; Fogagnolo, J.B. Microstructure and mechanical behavior of porous Ti-6Al-4V parts obtained by selective laser melting. *J. Mech. Behav. Biomed.* **2013**, *26*, 98–108. [CrossRef] [PubMed]

14. Shunmugavel, M.; Polishetty, A.; Littlefair, G. Microstructure and Mechanical Properties of Wrought and Additive Manufactured Ti-6Al-4V Cylindrical Bars. *Proc. Tech.* **2015**, *20*, 231–236. [CrossRef]

15. Murr, L.E.; Quinones, S.A.; Gaytan, S.M.; Lopez, M.I.; Rodela, A.; Martinez, E.Y.; Wicker, R.B. Microstructure and mechanical behavior of Ti-6Al-4V produced by rapid-layer manufacturing, for biomedical applications. *J. Mech. Behav. Biomed. Mater.* **2009**, *2*, 20–32. [CrossRef] [PubMed]

16. Tao, P.; Zhong, J.; Li, H.; Hu, Q.; Gong, S.; Xu, Q. Microstructure, Mechanical Properties, and Constitutive Models for Ti-6Al-4V Alloy Fabricated by Selective Laser Melting (SLM). *Metals* **2019**, *9*, 447. [CrossRef]

17. Greitemeier, D.; Dalle Donne, C.; Syassen, F.; Eufinger, J.; Melz, T. Effect of surface roughness on fatigue performance of additive manufactured Ti-6Al-4V. *Mater. Sci. Tech. Ser.* **2016**, *32*, 629–634. [CrossRef]

18. Zhang, Y.; Bernard, A.; Harik, R.; Karunakaran, K.P. Build orientation optimization for multi-part production in additive manufacturing. *J. Intell. Manuf.* **2017**, *28*, 1393–1407. [CrossRef]

19. Wauthle, R.; Vrancken, B.; Beynaerts, B.; Jorissen, K.; Schrooten, J.; Kruth, J.P.; Van Humbeeck, J. Effects of build orientation and heat treatment on the microstructure and mechanical properties of selective laser melted Ti-6Al-4V lattice structures. *Add. Manuf.* **2015**, *5*, 77–84.

20. Carroll, B.E.; Palmer, T.A.; Beese, A.M. Anisotropic tensile behavior of Ti-6Al-4V components fabricated with directed energy deposition additive manufacturing. *Acta Mater.* **2015**, *87*, 309–320. [CrossRef]

21. Mezzetta, J.; Choi, J.P.; Milligan, J.; Danovitch, J.; Chekir, N.; Bois-Brochu, A.; Brochu, M. Microstructure-Properties Relationships of Ti-6Al-4V Parts Fabricated by Selective Laser Melting. *Int. J. Precis. Eng. Manuf. Green Technol.* **2018**, *5*, 605–612. [CrossRef]

22. Li, P.; Warner, D.H.; Fatemi, A.; Phan, N. Critical assessment of the fatigue performance of additively manufactured Ti-6Al-4V and perspective for future research. *Int. J. Fatigue* **2016**, *85*, 130–143. [CrossRef]

23. Patterson, A.E.; Messimer, S.L.; Farrington, P.A. Overhanging features and the SLM/DMLS residual stresses problem: Review and future research need. *Technologies* **2017**, *5*, 15. [CrossRef]

24. Vrancken, B.; Thijs, L.; Kruth, J.P.; Van Humbeeck, J. Heat treatment of Ti-6Al-4V produced by Selective Laser Melting: Microstructure and mechanical properties. *J. Alloys Compd.* **2012**, *541*, 177–185. [CrossRef]

25. Vilaro, T.; Colin, C.; Bartout, J.D. As-fabricated and heat-treated microstructures of the Ti-6Al-4V alloy processed by selective laser melting. *Metall. Mater. Trans. A* **2014**, *42*, 3190–3199. [CrossRef]

26. Liu, Q.C.; Elambasseril, J.; Sun, S.J.; Leary, M.; Brandt, M.; Sharp, P.K. The effect of manufacturing defects on the fatigue behavior of Ti-6Al-4V specimens fabricated using selective laser melting. *Adv. Mat. Resh.* **2014**, *891*, 1519–1524.

27. Prabhu, A.W.; Vincent, T.; Chaudhary, A.; Zhang, W.; Babu, S.S. Effect of microstructure and defects on fatigue behavior of directed energy deposited Ti-6Al-4V. *Sci. Technol. Weld. Joi.* **2015**, *20*, 659–669. [CrossRef]

28. Razavi, S.M.J.; Bordonaro, G.G.; Ferro, P.; Torgersen, J.; Berto, F. Porosity effect on tensile behavior of Ti-6Al-4V specimens produced by laser engineered net shaping technology. *Proc. Inst. Mech. Eng. Part C: J. Mech. Eng. Sci.* **2018**. [CrossRef]

29. Gong, H.; Rafi, K.; Gu, H.; Ram, G.J.; Starr, T.; Stucker, B. Influence of defects on mechanical properties of Ti-6Al-4V components produced by selective laser melting and electron beam melting. *Mater. Des.* **2015**, *86*, 545–554. [CrossRef]

30. Zhang, G.; Wang, J.; Zhang, H. Research Progress of Balling Phenomena in Selective Laser Melting. *Found. Technol.* **2017**, *2*, 2.

31. Kasperovich, G.; Hausmann, J. Improvement of fatigue resistance and ductility of TiAl6V4 processed by selective laser melting. *J. Mater. Process. Tech.* **2015**, *220*, 202–214. [CrossRef]

32. Aboulkhair, N.T.; Everitt, N.M.; Ashcroft, I.; Tuck, C. Reducing porosity in AlSi10Mg parts processed by selective laser melting. *Add. Manuf.* **2014**, *1*, 77–86. [CrossRef]

33. Thijs, L.; Verhaeghe, F.; Craeghs, T.; Van Humbeeck, J.; Kruth, J.P. A study of the microstructural evolution during selective laser melting of Ti-6Al-4V. *Acta Mater.* **2010**, *58*, 3303–3312. [CrossRef]

34. Weingarten, C.; Buchbinder, D.; Pirch, N.; Meiners, W.; Wissenbach, K.; Poprawe, R. Formation and reduction of hydrogen porosity during selective laser melting of AlSi10Mg. *J. Mater. Process. Tech.* **2015**, *221*, 112–120. [CrossRef]

35. Brånemark, P.I. Osseointegration and its experimental background. *J. Prosth. Dent.* **1983**, *50*, 399–410. [CrossRef]

36. Rack, H.J.; Qazi, J.I. Titanium alloys for biomedical applications. *Mater. Sci. Eng. C* **2006**, *26*, 1269–1277. [CrossRef]

37. Wang, J.L.; Liu, R.L.; Majumdar, T.; Mantri, S.A.; Ravi, V.A.; Banerjee, R.; Birbilis, N. A closer look at the in vitro electrochemical characterization of titanium alloys for biomedical applications using in-situ methods. *Acta Biomater.* **2017**, *54*, 469–478. [CrossRef] [PubMed]

38. Harun, W.S.W.; Kamariah, M.S.I.N.; Muhamad, N.; Ghani, S.A.C.; Ahmad, F.; Mohamed, Z. A review of powder additive manufacturing processes for metallic biomaterials. *Powder Technol.* **2018**, *327*, 128–151. [CrossRef]

39. Karlsson, J.; Norell, M.; Ackelid, U.; Engqvist, H.; Lausmaa, J. Surface oxidation behavior of Ti-6Al-4V manufactured by Electron Beam Melting (EBM®). *J. Manuf. Process.* **2015**, *17*, 120–126. [CrossRef]

40. Du Plessis, A.; Kouprianoff, D.P.; Yadroitsava, I.; Yadroitsev, I. Mechanical Properties and In Situ Deformation Imaging of Microlattices Manufactured by Laser Based Powder Bed Fusion. *Materials* **2018**, *11*, 1663. [CrossRef]

41. Maamoun, A.; Xue, Y.; Elbestawi, M.; Veldhuis, S. Effect of selective laser melting process parameters on the quality of al alloy parts: Powder characterization, density, surface roughness, and dimensional accuracy. *Materials* **2018**, *11*, 2343. [CrossRef] [PubMed]

42. Maamoun, A.; Xue, Y.; Elbestawi, M.; Veldhuis, S. The Effect of Selective Laser Melting Process Parameters on the Microstructure and Mechanical Properties of Al6061 and AlSi10Mg Alloys. *Materials* **2019**, *12*, 12. [CrossRef] [PubMed]

43. Yu, M.; Yi, J.; Liu, J.; Li, S.; Wu, G.; Wu, L. Effect of electropolishing on electrochemical behaviours of titanium alloy Ti-10V-2Fe-3Al. *J. Wuhan Univ. Technol. Mater. Sci. Ed.* **2011**, *26*, 469–477. [CrossRef]

44. Han, J.; Yang, J.; Yu, H.; Yin, J.; Gao, M.; Wang, Z.; Zeng, X. Microstructure and mechanical property of selective laser melted Ti6Al4V dependence on laser energy density. *Rapid Prototyp. J.* **2017**, *23*, 217–226. [CrossRef]

45. Laquai, R.; Müller, B.R.; Kasperovich, G.; Haubrich, J.; Requena, G.; Bruno, G. X-ray refraction distinguishes unprocessed powder from empty pores in selective laser melting Ti-6Al-4V. *Mater. Res. Lett.* **2018**, *6*, 130–135. [CrossRef]

46. Dilip, J.J.S.; Zhang, S.; Teng, C.; Zeng, K.; Robinson, C.; Pal, D.; Stucker, B. Influence of processing parameters on the evolution of melt pool, porosity, and microstructures in Ti-6Al-4V alloy parts fabricated by selective laser melting. *Progress Add. Manuf.* **2017**, *2*, 157–167. [CrossRef]

47. Singh, A.P.; Yang, F.; Torrens, R.; Gabbitas, B. Solution treatment of Ti-6Al-4V alloy produced by consolidating blended powder mixture using a powder compact extrusion route. *Mater. Sci. Eng. A* **2018**, *712*, 157–165. [CrossRef]

48. Brezinová, J.; Hudák, R.; Guzanová, A.; Draganovská, D.; Ižaríková, G.; Koncz, J. Direct Metal Laser Sintering of Ti6Al4V for Biomedical Applications: Microstructure, Corrosion Properties and Mechanical Treatment of Implants. *Metals* **2016**, *6*, 171. [CrossRef]

49. Zhao, X.; Li, S.; Zhang, M.; Liu, Y.; Sercombe, T.B.; Wang, S.; Murr, L.E. Comparison of the microstructures and mechanical properties of Ti-6Al-4V fabricated by selective laser melting and electron beam melting. *Mater. Des.* **2016**, *95*, 21–31. [CrossRef]

50. Fan, Z.; Feng, H. Study on selective laser melting and heat treatment of Ti-6Al-4V alloy. *Results Phys.* **2018**, *10*, 660–664. [CrossRef]

51. Wang, D.; Dou, W.; Yang, Y. Research on selective laser melting of Ti6Al4V: Surface morphologies, optimized processing zone, and ductility improvement mechanism. *Metals* **2018**, *8*, 471. [CrossRef]

52. Dai, N.; Zhang, L.C.; Zhang, J.; Chen, Q.; Wu, M. Corrosion behavior of selective laser melted Ti-6Al-4V alloy in NaCl solution. *Corros. Sci.* **2016**, *102*, 484–489. [CrossRef]

53. Dai, N.; Zhang, J.; Chen, Y.; Zhang, L.C. Heat treatment degrading the corrosion resistance of selective laser melted Ti-6Al-4V alloy. *J. Electrochem. Soc.* **2017**, *164*, 428–434. [CrossRef]

Article

Biomechanical Effect of UHMWPE and CFR-PEEK Insert on Tibial Component in Unicompartmental Knee Replacement in Different Varus and Valgus Alignments

Yong-Gon Koh [1], Hyoung-Taek Hong [2] and Kyoung-Tak Kang [2,*]

[1] Joint Reconstruction Center, Department of Orthopaedic Surgery, Yonsei Sarang Hospital, 10 Hyoryeong-ro, Seocho-gu, Seoul 06698, Korea; osygkoh@gmail.com
[2] Department of Mechanical Engineering, Yonsei University, 50 Yonsei-ro, Seodaemun-gu, Seoul 03722, Korea; hyoungtaekhong@gmail.com
* Correspondence: tagi1024@gmail.com

Received: 1 August 2019; Accepted: 8 October 2019; Published: 14 October 2019

check for updates

Abstract: The current study aims to analyze the biomechanical effects of ultra-high molecular weight polyethylene (UHMWPE) and carbon-fiber-reinforced polyetheretherketone (CFR-PEEK) inserts, in varus/valgus alignment, for a tibial component, from 9° varus to 9° valgus, in unicompartmental knee replacement (UKR). The effects on bone stress, collateral ligament force, and contact stress on other compartments were evaluated under gait cycle conditions, by using a validated finite element model. In the UHMWPE model, the von Mises' stress on the cortical bone region significantly increased as the tibial tray was in valgus >6°, which might increase the risk of residual pain, and when in valgus >3° for CFR-PEEK. The contact stress on other UHMWPE compartments decreased in valgus and increased in varus, as compared to the neutral position. In CFR-PEEK, it increased in valgus and decreased in varus. The forces on medial collateral ligaments increased in valgus, when compared to the neutral position in UHMWPE and CFR-PEEK. The results indicate that UKR with UHMWPE showed positive biomechanical outputs under neutral and 3° varus conditions. UKR with CFR-PEEK showed positive biomechanical outputs for up to 6° varus alignments. The valgus alignment should be avoided.

Keywords: computational model; biomechanics; unicompartmental knee replacement; UHMWPE; CFR-PEEK; varus and valgus alignments

1. Introduction

Unicompartmental knee replacement (UKR) has become a popular alternative to total knee replacement (TKR), owing to favorable patient satisfaction reports and functional outcomes [1,2]. With a more functional anatomy being maintained, UKR can provide faster recovery and better restoration of knee kinetics, in comparison to TKR [3]. The long-term survival rates of UKR have been considerably improved, owing to refined surgical techniques and strict patient selection [4].

Various studies have shown that inaccurate alignment or technical errors in prosthetic components might cause early wear of polyethylene, periprosthetic fractures, and high revision rates to correct residual pain that is caused by implant failure [5–7]. The main complications are caused by abnormal tibial bone stress, which is thought to result from coronal malalignments of the tibial component. Therefore, UKR can be considered as a more technically demanding surgical treatment. In addition, numerous authors have suggested the relative malalignment of the knee in medial UKR, to avoid progressive osteoarthritis (OA) in the opposite compartment [8–10]. The effects of knee alignment on the long-term outcome of UKR have received attention in some reports [10,11].

The wear of ultra-high molecular weight polyethylene (UHMWPE), in both fixed- and mobile-bearings, continues to be a significant factor in the long-term performance of UKR [12]. Alternative polymers to replace UHMWPE in UKR or TKR have been proposed and they include the use of carbon-fiber reinforced polyetheretherketone (CFR-PEEK). In most cases, the carbon fibers in the CFR-PEEK materials are small, chopped fibers, and generally have no specific orientation in the polyetheretherketone (PEEK) matrix. CFR-PEEK has been classified as a low wear material in a recent in-vitro experimental study, being suitable for high-conformity scenarios, such as total hip replacement and conforming UKR [13]. However, it was recently stated that the wear rates of CFR-PEEK were very high and almost two orders of magnitude higher than the wear rate of ultra-high molecular weight polyethylene (UHMWPE) under comparable conditions for total knee replacement (TKR) [14]. S. C. Scholes and A. Unsworth stated that UKR with CFR-PEEK exhibited lower volumetric wear rates than those that were previously found in conventional metal-on-UHMWPE prostheses when tested under similar conditions [15]. Further study is required to confirm CFR-PEEK as an alternative material to UHMWPE. However, to the best of our knowledge, there has been a lack of research on the tibia bone stress when using CFR-PEEK tibial insert materials, which is an important factor determining aseptic loosening and anteromedial pain.

Most frequently, UKR is performed at a perpendicular coronal tibial position. However, various studies have reported improved results upon placing the tibial component parallel to the line of natural tibial joint, or within a few varus degrees of the epiphyseal axis [16,17]. There has been no study on the biomechanical effects of CFR-PEEK when compared to UHMWPE, with respect to varus and valgus alignments in the tibial component. Moreover, as mentioned above, there has been no biomechanical justification for the UHMWPE insert in the varus and valgus alignments in the tibial component.

The purpose of this study is to evaluate the biomechanical effects of the various varus/valgus alignment positions of tibial components in UKR with different UHMWPE and CFR-PEEK insert materials. We evaluate the contact stress on other compartments, the bone stress, and collateral ligament force, under a gait cycle condition, by using models from 9° varus to 9° valgus, with an increment of 1°, which is centered on the neutral position.

2. Materials and Methods

2.1. Intact Knee Joint Model

The present study was conducted while using a fully validated finite element (FE) model of knee joint [18–21]. The intact knee was made from computed tomography (CT) and magnetic resonance imaging (MRI) of the left knee joint of a healthy 35-year-old male volunteer. The CT imaging was conducted with a slice thickness of 0.1 mm, while using a 64-channel CT scanner (Somatom Sensation 64, Siemens Healthcare, Erlangen, Germany). In addition, MRI imaging was conducted using a 3.0 T MRI scanner (Achieva 3.0T; Philips Healthcare, North Brabant, The Netherlands) and a custom-designed knee joint cardiac coil [21]. MRI scans with a slice thickness of 0.4 mm were obtained in the sagittal plane. A high-resolution environment was used for the spectral pre-saturation inversion recovery (SPIR) sequence (TE, 25.0 ms; TR, 3590.8 ms; acquisition matrix, 512 pixels × 512 pixels; NEX, 2.0; field-of-view, 140 mm × 140 mm) and a 3-T MRI system (Discovery MR750w®, GE Healthcare, Milwaukee, WI, USA) [21].

The process of combining the reconstructed CT and MRI models through the positional alignment of each model was conducted while using the commercial software Rapid form (version 2006; 3D Systems Korea, Seoul, Korea). Subsequently, the image data were imported into the image-processing software Mimics 17.0 (Materialise Ltd., Leuven, Belgium) to extract the geometry used for the generation of the three-dimensional (3D) models of all structures. The initial graphics exchange specification (IGES) files that were exported from Mimics were entered into Unigraphics NX (version 7.0; Siemens

PLM Software, Torrance, CA, USA) to form solid models for each femur, tibia, fibula, patella, and soft-tissue segment (Figure 1). The solid model was imported into Hypermesh (version 8.0; Altair Engineering, Troy, MI, USA) to generate an FE mesh. Afterwards, the FE mesh was analyzed while using the ABAQUS software (version 6.11; Simulia, Providence, RI, USA).

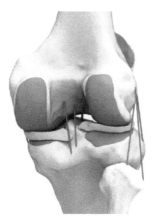

Figure 1. Developed three-dimensional finite element (3D FE) knee joint model from medical image.

In the model, the bones were assumed rigid, whereas the tibia was considered transversely isotropic as it was stiffer than soft tissue and had a minimal influence in the present study [22]. Constitutive laws were assumed for the tibia cortical and cancellous bones [23,24]. The cortical bone was considered to be transversely isotropic. The third axis was considered as parallel to the anatomical axis of each bone. The cancellous bone was considered as linear isotropic [23,24]. The articular cartilage was assumed to be an isotropic, linear elastic material, owing to the time-independent and simple compressive load applied to the knee joint [25]. The menisci were assumed to be a transversely isotropic, linearly elastic, homogeneous material [26]. The applied material properties were as shown in Table 1.

Table 1. Material properties for the 3D FE knee joint model.

Material	Material Properties
Cortical bone [23,24]	$E_1 = E_2 = 11.5\,\text{GPa}$ $E_3 = 17\,\text{GPa}$ $v_{12} = 0.51,$ $v_{23} = v_{13} = 0.31$
Cancellous bone [23,24]	$E = 2.13\,\text{GPa}$ $v = 0.3$
Articular cartilage [25]	$E = 15\,\text{MPa}$ $v = 0.47$
Menisci [26]	$E_c = 120\,\text{MPa}$ (in the circumferential direction) $E_a = E_r = 20\,\text{MPa MPa}$ (in the axial and radial directions.) $v_{cr} = 0.3$ (for axial directions) $v_{ar} = v_{ac} = 0.3$ (for circumferential and radial directions)

To represent the meniscal attachments, each meniscal horn was bonded to the bone using linear spring elements (SPRINGA element type) with a total stiffness of 2000 N/mm [27]. In addition, the major ligaments were modeled while using nonlinear and tension-only spring elements [28,29].

The force-displacement relationship, based on the functional bundles in the actual ligament anatomy, is expressed in the following:

$$f(\varepsilon) = \begin{cases} \frac{k\varepsilon^2}{4\varepsilon_1}, & 0 \leq \varepsilon \leq 2\varepsilon_1 \\ k(\varepsilon - \varepsilon_1), & \varepsilon > 2\varepsilon_1 \\ 0, & \varepsilon < 0 \end{cases}$$

$$\varepsilon = \frac{l - l_0}{l_0}$$

$$l_0 = \frac{l_r}{\varepsilon_r + 1}$$

where $f(\varepsilon)$ is the current force, k the stiffness, ε the strain, and ε_1 is assumed to be a constant equal to 0.03. The ligament bundle slack of length l_0 was calculated while using the reference bundle length l_r and the reference strain ε_r in the upright reference position.

The interfaces between the cartilage and bones were modeled as being fully bonded. The contacts between the femoral cartilage and meniscus, meniscus and tibial cartilage, and femoral cartilage and tibial cartilage were modeled for both the medial and lateral sides, and resulted in six contact pairs [18]. The contacts at all articulations adopted a finite sliding, frictionless hard contact algorithm with no penetration [18]. Convergence was defined as a relative change of more than 5% between the two adjacent meshes [20].

2.2. UKR Model

A fixed-bearing UKR (Zimmer, Inc., Warsaw, IN, USA) was virtually implanted in the medial compartment of the developed normal knee model. The bone models were imported and appropriately positioned, trimmed, and meshed with rigid elements by using surgical techniques [30].

Based on the dimensions of the femur and tibia, size 6 and 5 devices were chosen for the femoral and tibial components, respectively. The devices were aligned with the mechanical axis and positioned at the medial edge of the tibia. The neutrally aligned tibial baseplate was defined as having a square (0°) inclination in the coronal plane with a 5° posterior slope. The rotating axis was defined as a line parallel to the lateral edge of the tibial component passing through the center of the femoral component peg. A neutral femoral component distal cut, which was perpendicular to the mechanical axis of the femur and parallel to the tibial cut, was reproduced. The 19 different cut surfaces of the medial tibial plateau were created with different inclinations in the coronal plane (from 9° varus to 9° valgus, in increments of 1°) (Figure 2).

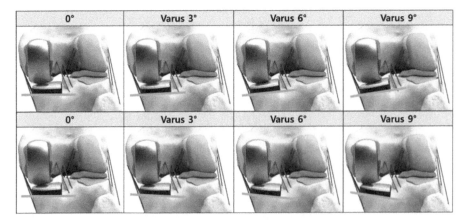

Figure 2. Schematic of varus/valgus conditions for tibial insert.

With respect to the implanted model, a cement gap of 1 mm was simulated between the bone and component. The femoral and tibial components, bone cement, and tibia insert (UHMWPE and CFR-PEEK) were modeled as linear elastic isotropic materials [31–35]. The materials of the femoral component, tibia insert, tibial component, and bone cement corresponded to a cobalt chromium alloy (CoCr), UHMWPE, CFR-PEEK, a titanium alloy (Ti6Al4V), and poly(methyl methacrylate) (PMMA), respectively. Compared to pure PEEK, CFR-PEEK contains a 30% proportion of carbon fiber reinforcement [35].

In terms of Young's modulus and Poisson's ratio, the material properties were shown in Table 2 [31–35]. The friction coefficients between the articulating surfaces were assumed to be 0.07 for UHMWPE and 0.04 for CFR-PEEK in accordance with the range that was reported in the literature [33,36,37].

Table 2. Material properties for the implanted model.

Materials	Material Properties
CoCr	E = 195 GPa v = 0.3
Ti6Al4V	E = 110 GPa v = 0.3
UHMWPE	E = 685 MPa v = 0.47
CFR-PEEK	E = 18000 MPa v = 0.4
PMMA	E = 1940 MPa and v = 0.4

2.3. Loading and Boundary Conditions

The FE investigation included two types of loading conditions, which corresponded to the loads that were used in the experimental part of the study, for the validation of the UKR model and its predictions under gait cycle loading conditions.

A validation of the intact model was conducted in a previous study, and the UKR model was validated through a comparison with models in previous experimental studies [18–21,38]. The validation of the UKR model was conducted for the flexion angles of 0°, 30°, 60°, and 90°, by using a passive flexion simulation. Additionally, anterior and posterior drawer loads of 134 N were separately applied to the tibia at the knee center, in a manner that is similar to that adopted in a previous experimental study [38].

Gait cycle loading was applied as a second loading to compare the biomechanical effects of the three different tibial insert materials [39]. A computational analysis was conducted while using the force controls of the anterior-posterior force with respect to the compressive load applied to the hip [40,41]. A proportional-integral derivative controller was incorporated into the computational model to make allowance for the control of the quadriceps, in a manner that is similar to that in an existing experiment [42]. Internal-external and varus/valgus torques were applied to the tibia [40,41]. We analyzed the contact stress on the lateral meniscus and tibial cartilage in the other compartments, as well as the tibial bone stress, which influences residual pain. Additionally, we analyzed the collateral ligament force. We defined five regions of interest (ROIs) on the proximal tibia (Figure 3). ROIs 1 and 4 were defined at the resection corner between the sagittal and transverse tibia bone cuts. The other three ROIs were located on the cancellous bone surface, beneath the tibial baseplate, with ROIs 2 and 4 being lateral and medial to the keel slot, respectively. ROI 5 was defined to investigate the source of residual pain. For all models, ROI 5 was located on the proximal anteromedial cortical bone surface.

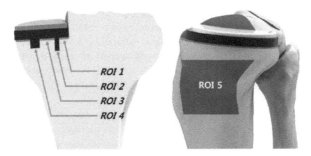

Figure 3. Five locations for the interests of this study.

Interventionary studies involving humans or animals, and other studies requiring ethical approval, must list the authority that provided approval and the corresponding ethical approval code.

3. Results

3.1. FE Model Validation

In a previous study, an intact model was validated and explained while using an in vivo laxity test for a subject identical to the FE model. We compared it with the results of the experiment on our own subject to validate the intact FE model. The anterior tibial translation was calculated at 2.83 mm in the experiment, and at 2.54 mm in the FE model under the 30° flexion loading condition; the posterior tibial translation was calculated at 2.12 mm in the experiment and 2.18 mm in the FE model for validation. Good agreement was observed between the experimental results and the FE model [19]. The UKR model validation was conducted while using the anterior and posterior tibial translations in the anterior and posterior drawer tests at 134 N for 6.1, 9.9, 8.7, and 8.5 mm, and 5.8, 4.3, 3.8, and 4.9 mm at 0°, 30°, 60°, and 90° knee flexion, respectively (Figure 4). These show good agreement with previous experimental data within the value ranges under the anterior and posterior drawer loadings [38].

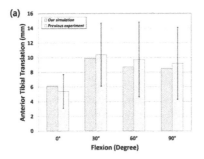
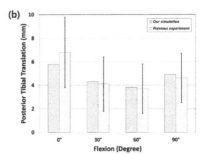

Figure 4. Comparison with previous experimental study for validation of unicompartmental knee replacement (UKR) model: (**a**) anterior tibial translation; and, (**b**) posterior tibial translation.

3.2. Comparison of Contact Stress on Other Compartments in UHMWPE and CFR-PEEK Tibial Component in Varus/Valgus Alignment

Figure 5 shows the contact stress on the other compartments (articular cartilage and lateral meniscus) in the UHMWPE and CFR-PEEK tibial insert with the varus/valgus tibial component. The contact stress on the articular cartilage and lateral meniscus increased in varus and decreased in valgus, in comparison to the neutral position in the UHMWPE insert. However, the contact stress on the articular cartilage and lateral meniscus decreased in varus and increased in valgus, in comparison to the neutral position of the CFR-PEEK insert. There was different contact stress during the stance phase in the UHMWPE and CFR-PEEK insert models. For the UHMWPE model, it increased by 11% and 8% for varus 9°, and decreased by 11% and 9% for valgus 9° in the lateral meniscus and articular cartilage, respectively, in comparison to the neutral position. For the CFR-PEEK model, it increased by 21% and 18% for valgus 9°, and decreased by 19% and 16% for varus 9° in the lateral meniscus and articular cartilage, respectively, in comparison to the neutral position.

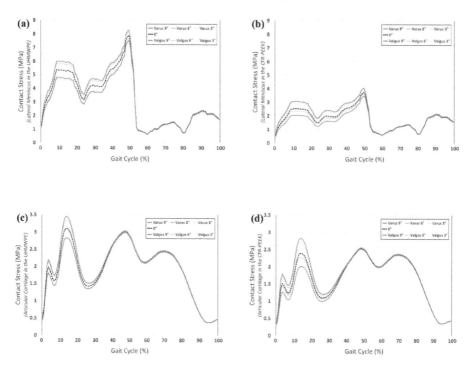

Figure 5. Comparison of contact stress on other compartments in varus/valgus condition with respect to different tibial insert materials during gait cycle: (**a**) lateral meniscus in the ultra-high molecular weight polyethylene (UHMWPE); (**b**) lateral meniscus in the carbon-fiber reinforced polyetheretherketone (CFR-PEEK); (**c**) articular cartilage in the UHMWPE; and, (**d**) articular cartilage in the CFR-PEEK.

3.3. Comparison of Von-Mises Stress in Tibial Bone and Collateral Ligament Force in UHMWPE and CFR-PEEK Tibial Component in Varus/Valgus Alignment

Figure 6 indicates the von-Mises stress on the tibia bone in the UHMWPE and CFR-PEEK tibial inserts with the varus/valgus tibial component. The greatest stress was exerted on ROI 1 and 2 (cancellous bone), in both the UHMWPE and CFR-PEEK. This trend did not change with varus and valgus in the tibial component. In addition, the bone stress in ROI 1–4, in the CFR-PEEK, was higher than in the UHMWPE. The von-Mises stress on the UHMWPE in the tibial bone showed a similar value from the neutral position to the varus 3° model. The CFR-PEEK model showed a similar value from the neutral position to varus 6°; however, it rapidly decreased afterwards, under the valgus condition. The stress in ROI 5 (cortical bone) with UHMWPE was greater than with CFR-PEEK. In addition, a similar trend was observed in ROI 1–4, for both UHMWPE and CFR-PEEK, under the varus and valgus conditions. Figure 7 indicates the force on the collateral ligament in the UHMWPE and CFR PEEK tibial insert with a varus/valgus tibial component. The force on the medial collateral ligament with the UHMWPE and CFR-PEEK increased under the valgus condition, in comparison to the neutral position. The force in the UHMWPE model with valgus 9° increased by 19% in the medial- collateral ligament, as compared to the neutral position. Moreover, the force on the medial-collateral ligament increased more in the CFR-PEEK model under the valgus condition. The force in the UHMWPE model with valgus 9° increased by 27% in the medial collateral ligament, in comparison to the neutral position. A complex pattern was observed in the lateral collateral ligament for the UHMWPE and CFR-PEEK. Under the valgus conditions, the force on the lateral collateral ligament increased during the swing and stance phases, respectively.

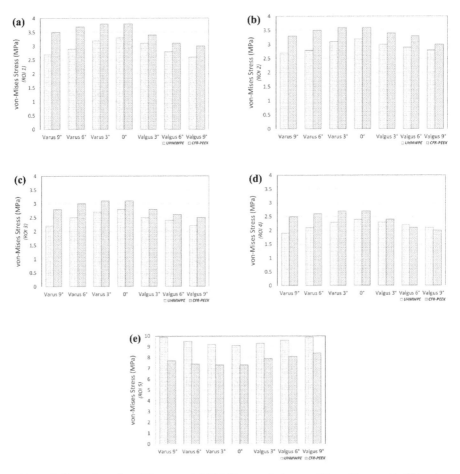

Figure 6. Comparison of von-Mises stress in ROI 1–5 in varus/valgus condition with respect to different tibial insert materials during gait cycle: (**a**) ROI 1; (**b**) ROI 2; (**c**) ROI 3; (**d**) ROI 4; and, (**e**) ROI 5.

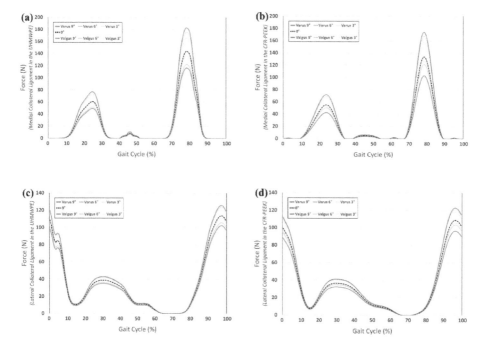

Figure 7. Comparison of ligament force in varus/valgus condition with respect to different tibial insert materials during gait cycle: (**a**) medial-collateral ligament in the UHMWPE; (**b**) medial-collateral ligament in the CFR-PEEK; (**c**) lateral-collateral ligament in the UHMWPE; and, (**d**) lateral-collateral ligament in the CFR-PEEK.

4. Discussion

The positive biomechanical effect that was observed in the UHMWPE tibial insert at the neutral position and with slight varus malalignment was the most important finding of the current study. Additionally, for the CFR-PEEK tibial insert, there was no problem with a varus malalignment greater than that of the UHMWPE tibial insert; however, the valgus malalignment should be avoided.

The precise restoration of the mechanical axis and correct implant positioning are major contributors to the improvement of implant longevity and the clinical outcomes of UKR. There is still no general agreement in terms of the optimal position of the tibial component [43].

This study aimed to evaluate the optimal tibial alignment for UKR with UHMWPE and CFR-PEEK tibial inserts, and the potential aberrant performances due to the malalignment of the tibial components. An FE model was used to analyze and compare different combinations of several biomechanical outputs. In addition, the FE model was impractical in the experimental evaluation of progressive OA in the other compartments and tibia bone stress problems, with respect to different insert materials under the varus/valgus conditions. Furthermore, the advantage of computational simulations that use a single subject is that the effects of tibial component alignment and the differences in insert materials, within the same subject, can be determined, except for variables, such as weight, height, bone geometry, differences in ethnicity and sex, ligament properties, and component size [44]. The intact knee model underwent a series of validation steps. The results showed good agreement with previous experimental data, in terms of kinematics and contact area, as demonstrated by the FE analysis for the same subject. In addition, an intact and a UKR model were both validated by using experimental and kinematics data. Therefore, the UKR model developed in the current study, along with the following analysis, can be considered to be reasonable.

As mentioned above, the UHMWPE insert led to result variation in the coronal positioning of the tibial tray. Sawatari et al. [45] analyzed the orientation influence of the unicompartmental tibial component on cancellous bone stresses in the coronal and sagittal planes, and correlated these findings to clinical data. They suggested that a slight valgus inclination of a UKR tibial component is better than a varus or square inclination in the coronal plane. In addition, Isekawa et al. [46] analyzed the inclination influence of the UKR tibial component on stress, in the proximal tibia, and the influence of contact pressure on the metal–bone interface, in the identical femorotibial alignment. They suggested that a slight valgus inclination of the tibial component might be better than varus or even a 0° (square) inclination, as the stress distribution is related. Therefore, a slight valgus inclination is preferable for both. Simpson et al. [47] found that the inclination angles had a minimal effect on the bone strain in the Oxford UKR, with the exclusion of the 2° varus inclination. However, there is a limitation in the three models that are mentioned above, as they only model tibia and not femur or other bony structure and soft tissues such as cartilage and ligament. Recently, Zhu et al. [48] developed a UKR model with entirely bony structure and soft tissue. They investigated the coronal inclination angles of the tibial tray ranging from 10° varus to 10° valgus. They suggested a range from 4° varus to 4° valgus for the inclination of the tibial component in the mobile-bearing UKR. In addition, Innocenti et al. [49] evaluated the biomechanical effects of different varus/valgus alignment tibial component positions in UKR. They proposed that a neutral mechanical or 3 varus alignment exhibits similar biomechanical outputs in the bone, collateral ligament strain, and polyethylene insert. A 6° varus alignment or changes in the valgus alignment were always associated with more detrimental effects. The two studies mentioned above also showed limitations, as they only evaluated the biomechanical effect under static conditions. However, UKR is a surgical treatment for younger patients; therefore, it requires analysis under dynamic loading conditions, such as gait cycles.

We studied the contact stress on other compartments, the stress on the tibial bone and the force on the collateral ligament under a gait cycle. Our results showed that alignment influences the development of OA in other compartments. Previous studies that have attempted to evaluate the influence of alignment were limited due to their short follow-up and absence of accurate knee alignment measurements [10,11,50]. In addition, a previous clinical study only used a radiography parameter (joint space narrowing) for the development of OA in the opposite compartment. This narrowing of joint space shows the articular cartilage loss in the joint and it is considered to be a more reliable marker of OA than the osteophytes or subchondral sclerosis. However, the radiography parameter might cause results to vary, owing to scanning inaccuracies and human error. We evaluated contact stress to investigate the progressive OA in other compartments because the contact parameters are known to be closely related to the degenerative OA of the knee joint [18,51]. An intriguing finding was that the effect of tibial malalignment was different for the different tibial insert materials. In UHMWPE, the contact stress on other compartments decreased in valgus and slightly increased under varus conditions [52]. This trend proved to be in fine agreement with a previous study. The contact force on the other compartment decreased under the valgus condition using UHMWPE and in the cadaveric experiment [52]. However, the CFR-PEEK model it increased in valgus and decreased in varus. Additionally, the difference between contact stresses on other compartments, in both the UHMWPE and CFR-PEEK model, was found during the stance phase. This proved that the contact stress was mainly influenced by the axial force.

The cortical bone always exerted the majority of the load transferred by the tibial baseplate, as its elastic modulus is much larger than that of the cancellous bone. An abnormally high stress of the proximal medial cortical bone was used to understand the cause of persistent pain after UKR [47]. Additionally, the aseptic loosening of prosthetic components, and especially that of the tibial component, is one of the major failure modes in UKR [53]. Excessive stress in both the cortical and cancellous bone cause it, and it leads to stress shielding. The major reason of the latter is the significant difference in the Young's modulus between the bone and tibial baseplate material. In contrast to TKR, the interface between the tibial baseplate and the tibia is significantly smaller. This leads to bone stresses being more sensitive to malalignment and malpositioning of component.

This UHMWPE model showed higher stress in the cortical bone region and lower stress in the cancellous bone region, in comparison to the CFR-PEEK model. The tibial insert material was only changed to CFR-PEEK; however, the stress-shielding effect was reduced because the material properties of CFR-PEEK were similar to the bone's material properties. The increase of bone stresses in the cortical bone and the decrease of bone stresses in the cancellous bone around the tibial baseplate may cause pain, which is induced by a loosening of the tibial baseplate [54–57]. Stress in the cortical bone increased, while, in the cancellous bone, it decreased under the valgus condition for both UHMWPE and CFR-PEEK, in comparison to the neutral position. In other words, the valgus malalignment should be avoided in both UHMWPE and CFR-PEEK.

After medial UKR was implanted, the valgus deformity was induced, even though the neutral knee model had a normal alignment (valgus rotation 0°) [49,58]. This phenomenon can be explained by the different stiffness between the medial and lateral compartments of the UKR knee. On the lateral side, the cartilage of both the femur and tibia had an elastic modulus of 15 MPa. In contrast to the medial side elastic moduli of the tibial insert materials UHMWPE and CFR-PEEK, the elastic modulus was much higher than that of the cartilage. Consequently, the Young's Modulus of the medial and lateral compartments differed by more than one order of magnitude; therefore, the deformation of materials in each compartment depended on their elastic modulus [54]. In a previous study, the valgus deformity that was induced by the strain loading, in the medial-collateral ligament, increased, while the strain in the lateral collateral ligament decreased [54]. Hence, whilst a surgeon balances a knee during UKR in an unloaded state, the knee will no longer be balanced once it is loaded [54].

Our study confirmed this result. The force on the medial-collateral ligament also increased for the UHMWPE and CFR-PEEK models under the valgus condition. This trend was particularly observed in CFR-PEEK. The reason was that CFR-PEEK has a higher elastic modulus than UHMWPE, as mentioned previously. Therefore, the difference in material properties became greater here than in other areas. Our results showed that progressive OA could occur in other compartments under the valgus condition. Additionally, valgus should be avoided because the MCL force rapidly increases in the CFR-PEEK model.

There are several strengths to this study. First, the FE model included the femur, tibia, and related soft tissues in the present study [31,45–47]. Second, gait cycle loading, and complex vertical static loading condition was applied in this study [31,45–49,54]. Third, this study validated not just the initial FE model, but also achieved kinematic validation of the UKR FE models with experimental data [31,45–49,54].

Nevertheless, there are three limitations. First, the FE model represented a fixed bearing UKR, and it cannot apply other implant designs, such as the mobile bearing UKR. Second, a single coefficient of friction value was used, when there is in fact a variety in the coefficients of friction for different materials. The effect of different values will be investigated in future studies.

Third, patient satisfaction and clinical results could not be assessed by the results from FE analysis. However, contact stress on the other compartments, force exerted on ligaments, and bone stresses are important factors that should be investigated for the evaluation of biomechanical effects in computational biomechanics [18–21,31,45–49].

5. Conclusions

The biomechanical justifications for the tibial insert materials UHMWPE and CFR PEEK were evaluated to confirm the suitability of various tibial component alignment positions while using an FE model. We performed a simulation for other compartments to investigate progressive OA. Contact stress on other compartments increased in varus and decreased in valgus for UHMWPE. In CFR-PEEK, an opposite trend was observed, by which the stress-shielding effect was less than that of UHMWPE, because its elastic modulus was closer to that of bone. However, in the CFR-PEEK model, a significant negative biomechanical effect was observed, owing to the valgus condition, in comparison to UHMWPE. In this model, the stress on the cortical and cancellous bone increased and decreased rapidly under the valgus and varus condition, respectively, in comparison to the neutral position. This trend was also observed in UHMWPE; however, the negative biomechanical effects were more pronounced in CFR PEEK; here, the varus allowance was wider than in UHMWPE, as there was less difference in the biomechanical effects from the neutral position to varus 6°. Such a trend could be found in the medial collateral ligament. The force on the medial collateral ligament became greater under the valgus condition of the CFR-PEEK model, in comparison to the UHMWPE model. The UHMPWE model, with a neutral position or minor varus (3°) alignment, exhibits similar biomechanical outputs in terms of the contact stress on other compartments, the stress on the bone, and the collateral ligament force. Therefore, the neutral position and minor varus conditions are recommended in the UHMWPE model. In addition, a greater varus condition of up to 6° is recommended in the CFR-PEEK model, as it showed similar biomechanical output. However, the valgus condition should be avoided in the CFR-PEEK model.

Author Contributions: Y.-G.K. and H.-T.H. contributed equally to the current study and they should be considered as co-first authors. K.-T.K.—Conceptualization; Validation; Writing—reviewing & editing; Methodology.

Funding: The current research did not receive an external funding.

Conflicts of Interest: The authors declare no conflict of interest.

References

1. Berger, R.A.; Meneghini, R.M.; Jacobs, J.J.; Sheinkop, M.B.; Della Valle, C.J.; Rosenberg, A.G.; Galante, J.O. Results of unicompartmental knee arthroplasty at a minimum of ten years of follow-up. *J. Bone Joint Surg. Am.* **2005**, *87*, 999–1006. [CrossRef] [PubMed]

2. Riddle, D.L.; Jiranek, W.A.; McGlynn, F.J. Yearly incidence of unicompartmental knee arthroplasty in the United States. *J. Arthroplast.* **2008**, *23*, 408–412. [CrossRef] [PubMed]

3. Mochizuki, T.; Sato, T.; Tanifuji, O.; Kobayashi, K.; Koga, Y.; Yamagiwa, H.; Omori, G.; Endo, N. In vivo pre- and postoperative three-dimensional knee kinematics in unicompartmental knee arthroplasty. *J. Orthop. Sci.* **2013**, *18*, 54–60. [CrossRef] [PubMed]

4. Pandit, H.; Jenkins, C.; Gill, H.S.; Barker, K.; Dodd, C.A.; Murray, D.W. Minimally invasive Oxford phase 3 unicompartmental knee replacement: Results of 1000 cases. *J. Bone Joint Surg. Br.* **2011**, *93*, 198–204. [CrossRef] [PubMed]

5. McAuley, J.P.; Engh, G.A.; Ammeen, D.J. Revision of failed unicompartmental knee arthroplasty. *Clin. Orthop. Relat. Res.* **2001**, 279–282. [CrossRef] [PubMed]

6. Weinstein, J.N.; Andriacchi, T.P.; Galante, J. Factors influencing walking and stairclimbing following unicompartmental knee arthroplasty. *J. Arthroplast.* **1986**, *1*, 109–115. [CrossRef]

7. Wilcox, P.G.; Jackson, D.W. Unicompartmental knee arthroplasty. *Orthop. Rev.* **1986**, *15*, 490–495.

8. Goodfellow, J.W.; Kershaw, C.J.; Benson, M.K.; O'Connor, J.J. The Oxford Knee for unicompartmental osteoarthritis. The first 103 cases. *J. Bone Joint Surg. Br.* **1988**, *70*, 692–701. [CrossRef]

9. Klemme, W.R.; Galvin, E.G.; Petersen, S.A. Unicompartmental knee arthroplasty. Sequential radiographic and scintigraphic imaging with an average five-year follow-up. *Clin. Orthop. Relat. Res.* **1994**, *301*, 233–238.

10. Weale, A.E.; Murray, D.W.; Baines, J.; Newman, J.H. Radiological changes five years after unicompartmental knee replacement. *J. Bone Joint Surg. Br.* **2000**, *82*, 996–1000. [CrossRef]

11. Kennedy, W.R.; White, R.P. Unicompartmental arthroplasty of the knee. Postoperative alignment and its influence on overall results. *Clin. Orthop. Relat. Res.* **1987**, *221*, 278–285.

12. Blunn, G.W.; Joshi, A.B.; Minns, R.J.; Lidgren, L.; Lilley, P.; Ryd, L.; Engelbrecht, E.; Walker, P.S. Wear in retrieved condylar knee arthroplasties. A comparison of wear in different designs of 280 retrieved condylar knee prostheses. *J. Arthroplast.* **1997**, *12*, 281–290. [CrossRef]

13. Brockett, C.L.; Carbone, S.; Fisher, J.; Jennings, L.M. *PEEK and CFR PEEK as an Alternative to UHMWPE in Total Knee Replacement*; ORS Annual Meeting Las Vegas: Las Vegas, NV, USA, 2015.

14. Brockett, C.L.; Carbone, S.; Fisher, J.; Jennings, L.M. PEEK and CFR-PEEK as alternative bearing materials to UHMWPE in a fixed bearing total knee replacement: An experimental wear study. *Wear* **2017**, *374–375*, 86–91. [CrossRef] [PubMed]

15. Scholes, S.C.; Unsworth, A. Pitch-based carbon-fibre-reinforced poly (ether-ether-ketone) OPTIMA assessed as a bearing material in a mobile bearing unicondylar knee joint. *Proc. Inst. Mech. Eng. H* **2009**, *223*, 13–25. [CrossRef] [PubMed]

16. Cartier, P.; Sanouiller, J.L.; Grelsamer, R.P. Unicompartmental knee arthroplasty surgery. 10-year minimum follow-up period. *J. Arthroplast.* **1996**, *11*, 782–788. [CrossRef]

17. Thornhill, T.S.; Scott, R.D. Unicompartmental total knee arthroplasty. *Orthop. Clin. N. Am.* **1989**, *20*, 245–256.

18. Kim, Y.S.; Kang, K.T.; Son, J.; Kwon, O.R.; Choi, Y.J.; Jo, S.B.; Choi, Y.W.; Koh, Y.G. Graft extrusion related to the position of allograft in lateral meniscal allograft transplantation: Biomechanical comparison between parapatellar and transpatellar approaches using finite element analysis. *Arthroscopy* **2015**, *31*, 2380–2391. [CrossRef]

19. Kang, K.T.; Kim, S.H.; Son, J.; Lee, Y.H.; Chun, H.J. Computational model-based probabilistic analysis of in vivo material properties for ligament stiffness using the laxity test and computed tomography. *J. Mater. Sci. Mater. Med.* **2016**, *27*, 183. [CrossRef]

20. Kwon, O.R.; Kang, K.T.; Son, J.; Suh, D.S.; Baek, C.; Koh, Y.G. Importance of joint line preservation in unicompartmental knee arthroplasty: Finite element analysis. *J. Orthop. Res.* **2017**, *35*, 347–352. [CrossRef]

21. Kang, K.T.; Kim, S.H.; Son, J.; Lee, Y.H.; Kim, S.; Chun, H.J. Probabilistic evaluation of the material properties of the in vivo subject-specific articular surface using a computational model. *J. Biomed. Mater. Res. B Appl. Biomater.* **2017**, *105*, 1390–1400. [CrossRef]

22. Pena, E.; Calvo, B.; Martinez, M.A.; Palanca, D.; Doblare, M. Why lateral meniscectomy is more dangerous than medial meniscectomy. A finite element study. *J. Orthop. Res.* **2006**, *24*, 1001–1010. [CrossRef] [PubMed]

23. Kayabasi, O.; Ekici, B. The effects of static, dynamic and fatigue behavior on three-dimensional shape optimization of hip prosthesis by finite element method. *Mater. Des.* **2007**, *28*, 2269–2277. [CrossRef]

24. Hoffler, C.E.; Moore, K.E.; Kozloff, K.; Zysset, P.K.; Goldstein, S.A. Age, gender, and bone lamellae elastic moduli. *J. Orthop. Res.* **2000**, *18*, 432–437. [CrossRef] [PubMed]

25. Shepherd, D.E.; Seedhom, B.B. The 'instantaneous' compressive modulus of human articular cartilage in joints of the lower limb. *Rheumatology* **1999**, *38*, 124–132. [CrossRef]

26. Haut Donahue, T.L.; Hull, M.L.; Rashid, M.M.; Jacobs, C.R. How the stiffness of meniscal attachments and meniscal material properties affect tibio-femoral contact pressure computed using a validated finite element model of the human knee joint. *J. Biomech.* **2003**, *36*, 19–34. [CrossRef]

27. Guess, T.M.; Thiagarajan, G.; Kia, M.; Mishra, M. A subject specific multibody model of the knee with menisci. *Med. Eng. Phys.* **2010**, *32*, 505–515. [CrossRef] [PubMed]

28. Takeda, Y.; Xerogeanes, J.W.; Livesay, G.A.; Fu, F.H.; Woo, S.L. Biomechanical function of the human anterior cruciate ligament. *Arthroscopy* **1994**, *10*, 140–147. [CrossRef]

29. Blankevoort, L.; Huiskes, R. Validation of a three-dimensional model of the knee. *J. Biomech.* **1996**, *29*, 955–961. [CrossRef]

30. Zimmer, I. *Zimmer Unicompartmental High Flex Knee: Intramedullary, Spacer Block Option and Extramedullary Minimally Invasive Surgical Techniques*; Zimmer Biomet: Warsaw, IN, USA, 2005.

31. Inoue, S.; Akagi, M.; Asada, S.; Mori, S.; Zaima, H.; Hashida, M. The Valgus Inclination of the Tibial Component Increases the Risk of Medial Tibial Condylar Fractures in Unicompartmental Knee Arthroplasty. *J. Arthroplast.* **2016**, *31*, 2025–2030. [CrossRef]

32. Pegg, E.C.; Walter, J.; Mellon, S.J.; Pandit, H.G.; Murray, D.W.; D'Lima, D.D.; Fregly, B.J.; Gill, H.S. Evaluation of factors affecting tibial bone strain after unicompartmental knee replacement. *J. Orthop. Res.* **2013**, *31*, 821–828. [CrossRef]

33. Godest, A.C.; Beaugonin, M.; Haug, E.; Taylor, M.; Gregson, P.J. Simulation of a knee joint replacement during a gait cycle using explicit finite element analysis. *J. Biomech.* **2002**, *35*, 267–275. [CrossRef]

34. Innocenti, B.; Truyens, E.; Labey, L.; Wong, P.; Victor, J.; Bellemans, J. Can medio-lateral baseplate position and load sharing induce asymptomatic local bone resorption of the proximal tibia? A finite element study. *J. Orthop. Surg. Res.* **2009**, *4*, 26. [CrossRef] [PubMed]

35. Kang, K.-T.; Koh, Y.-G.; Son, J.; Yeom, J.S.; Park, J.-H.; Kim, H.-J. Biomechanical evaluation of pedicle screw fixation system in spinal adjacent levels using polyetheretherketone, carbon-fiber-reinforced polyetheretherketone, and traditional titanium as rod materials. *Compos. Part B Eng.* **2017**, *130*, 248–256. [CrossRef]

36. Knight, L.A.; Pal, S.; Coleman, J.C.; Bronson, F.; Haider, H.; Levine, D.L.; Taylor, M.; Rullkoetter, P.J. Comparison of long-term numerical and experimental total knee replacement wear during simulated gait loading. *J. Biomech.* **2007**, *40*, 1550–1558. [CrossRef] [PubMed]

37. Greco, A.C.; Erck, R.; Ajayi, O.; Fenske, G. Effect of reinforcement morphology on high-speed sliding friction and wear of PEEK polymers. *Wear* **2011**, *271*, 2222–2229. [CrossRef]

38. Suggs, J.F.; Li, G.; Park, S.E.; Steffensmeier, S.; Rubash, H.E.; Freiberg, A.A. Function of the anterior cruciate ligament after unicompartmental knee arthroplasty: An in vitro robotic study. *J. Arthroplast.* **2004**, *19*, 224–229. [CrossRef] [PubMed]

39. Standardization, I.O. f. *ISO 14243-1: Implants for Surgery—Wear of Total Knee-Joint Prostheses—Part 1: Loading and Displacement Parameters for Wear-Testing Machines with Load Control and Corresponding Environmental Conditions for Test*; International Organization for Standardization: Geneva, Switzerland, 2002.

40. Halloran, J.P.; Clary, C.W.; Maletsky, L.P.; Taylor, M.; Petrella, A.J.; Rullkoetter, P.J. Verification of predicted knee replacement kinematics during simulated gait in the Kansas knee simulator. *J. Biomech. Eng.* **2010**, *132*, 081010. [CrossRef]

41. Kutzner, I.; Heinlein, B.; Graichen, F.; Bender, A.; Rohlmann, A.; Halder, A.; Beier, A.; Bergmann, G. Loading of the knee joint during activities of daily living measured in vivo in five subjects. *J. Biomech.* **2010**, *43*, 2164–2173. [CrossRef]

42. Kang, K.T.; Koh, Y.G.; Son, J.; Kwon, O.R.; Baek, C.; Jung, S.H.; Park, K.K. Measuring the effect of femoral malrotation on knee joint biomechanics for total knee arthroplasty using computational simulation. *Bone Joint Res.* **2016**, *5*, 552–559. [CrossRef]

43. Keene, G.; Simpson, D.; Kalairajah, Y. Limb alignment in computer-assisted minimally-invasive unicompartmental knee replacement. *J. Bone Joint Surg. Br.* **2006**, *88*, 44–48. [CrossRef]

44. Thompson, J.A.; Hast, M.W.; Granger, J.F.; Piazza, S.J.; Siston, R.A. Biomechanical effects of total knee arthroplasty component malrotation: A computational simulation. *J. Orthop. Res.* **2011**, *29*, 969–975. [CrossRef] [PubMed]

45. Sawatari, T.; Tsumura, H.; Iesaka, K.; Furushiro, Y.; Torisu, T. Three-dimensional finite element analysis of unicompartmental knee arthroplasty—The influence of tibial component inclination. *J. Orthop. Res.* **2005**, *23*, 549–554. [CrossRef] [PubMed]

46. Iesaka, K.; Tsumura, H.; Sonoda, H.; Sawatari, T.; Takasita, M.; Torisu, T. The effects of tibial component inclination on bone stress after unicompartmental knee arthroplasty. *J. Biomech.* **2002**, *35*, 969–974. [CrossRef]

47. Simpson, D.J.; Price, A.J.; Gulati, A.; Murray, D.W.; Gill, H.S. Elevated proximal tibial strains following unicompartmental knee replacement—A possible cause of pain. *Med. Eng. Phys.* **2009**, *31*, 752–757. [CrossRef] [PubMed]

48. Zhu, G.D.; Guo, W.S.; Zhang, Q.D.; Liu, Z.H.; Cheng, L.M. Finite Element Analysis of Mobile-bearing Unicompartmental Knee Arthroplasty: The Influence of Tibial Component Coronal Alignment. *Chin. Med. J.* **2015**, *128*, 2873–2878. [CrossRef] [PubMed]

49. Innocenti, B.; Pianigiani, S.; Ramundo, G.; Thienpont, E. Biomechanical Effects of Different Varus and Valgus Alignments in Medial Unicompartmental Knee Arthroplasty. *J. Arthroplast.* **2016**, *31*, 2685–2691. [CrossRef]

50. Hernigou, P.; Deschamps, G. Alignment influences wear in the knee after medial unicompartmental arthroplasty. *Clin. Orthop. Relat. Res.* **2004**, *423*, 161–165. [CrossRef]

51. Segal, N.A.; Anderson, D.D.; Iyer, K.S.; Baker, J.; Torner, J.C.; Lynch, J.A.; Felson, D.T.; Lewis, C.E.; Brown, T.D. Baseline articular contact stress levels predict incident symptomatic knee osteoarthritis development in the MOST cohort. *J. Orthop. Res.* **2009**, *27*, 1562–1568. [CrossRef]

52. Heyse, T.J.; El-Zayat, B.F.; De Corte, R.; Scheys, L.; Chevalier, Y.; Fuchs-Winkelmann, S.; Labey, L. Balancing UKA: Overstuffing leads to high medial collateral ligament strains. *Knee Surg. Sports Traumatol. Arthrosc.* **2016**, *24*, 3218–3228. [CrossRef]

53. Berger, R.A.; Nedeff, D.D.; Barden, R.M.; Sheinkop, M.M.; Jacobs, J.J.; Rosenberg, A.G.; Galante, J.O. Unicompartmental knee arthroplasty. Clinical experience at 6- to 10-year followup. *Clin. Orthop. Relat. Res.* **1999**, *367*, 50–60.

54. Innocenti, B.; Bilgen, O.F.; Labey, L.; van Lenthe, G.H.; Sloten, J.V.; Catani, F. Load sharing and ligament strains in balanced, overstuffed and understuffed UKA. A validated finite element analysis. *J. Arthroplast.* **2014**, *29*, 1491–1498. [CrossRef] [PubMed]

55. Bartel, D.L.; Burstein, A.H.; Santavicca, E.A.; Insall, J.N. Performance of the tibial component in total knee replacement. *J. Bone Joint. Surg. Am.* **1982**, *64*, 1026–1033. [CrossRef] [PubMed]

56. Bourne, R.B.; Finlay, J.B. The influence of tibial component intramedullary stems and implant-cortex contact on the strain distribution of the proximal tibia following total knee arthroplasty. An in vitro study. *Clin. Orthop. Relat. Res.* **1986**, *208*, 95–99. [CrossRef]

57. Taylor, M.; Tanner, K.E.; Freeman, M.A. Finite element analysis of the implanted proximal tibia: A relationship between the initial cancellous bone stresses and implant migration. *J. Biomech.* **1998**, *31*, 303–310. [CrossRef]

58. Au, A.G.; Raso, V.J.; Liggins, A.B.; Otto, D.D.; Amirfazli, A. A three-dimensional finite element stress analysis for tunnel placement and buttons in anterior cruciate ligament reconstructions. *J. Biomech.* **2005**, *38*, 827–832. [CrossRef]

 *materials*

Article

Comparison of Titanium and Bioresorbable Plates in "A" Shape Plate Properties—Finite Element Analysis

Rafał Zieliński [1,*], Marcin Kozakiewicz [1] and Jacek Świniarski [2]

[1] Department of Maxillofacial Surgery, Medical University of Lodz, 1st Haller Plac, 90-647 Lodz, Poland; qed@op.pl

[2] Department of Strength of Materials and Structures, Technical University of Lodz, Stefanowskiego 1/15, 90-924 Lodz, Poland; jacek.swiniarski@p.lodz.pl

* Correspondence: bkost@op.pl

Received: 30 November 2018; Accepted: 21 March 2019; Published: 3 April 2019

Abstract: (1) Background: The main disadvantage of rigid fracture fixation is remain material after healing period. Implementation of resorbable plates prevents issues resulting from left plates. The aim of this study is to compare the usage of bioresorbable and titanium "A" shape condyle plate in condylar fractures. (2) Methods: Thickness of 1.0 mm, height of 31 mm, and width of 19 mm polylactic acid (PLLA) and titanium "A" shape plate with 2.0 mm-wide connecting bar and 9 holes were tested with finite element analysis in high right condylar neck fracture. (3) Results: On bone surface the highest stress is on the anterior bridge around first hole (approx. 100 MPa). The highest stress on screws is located in the first screw around plate in the anterior bridge and is greater in titanium (150 MPa) than PLLA (114 MPa). (4) Conclusion: Pressure on bone in PLLA osteosynthesis is two times higher than in titanium fixation. On small areas where pressure on bone is too high it causes local bone degradation around the fracture and may delay the healing process or make it impossible. Fixation by PLLA is such flexible that bone edges slide and twist what may lead to degradation of callus.

Keywords: mandible condylar fractures; surgical treatment; titanium; PLLA; finite element analysis

1. Introduction

Open reduction internal fixation is the method of choice in mandibular condyle fractures [1–3]. Although it demands the surgical skills because of the risk of facial palsy. Titanium alloys are still gold standard in terms of osteosynthesis fixation systems, however, bioresorbable plates and screws are also used in condyle fractures. The main disadvantage of metal plates is when the necessity of removal of osteosynthesis material appears (especially extraoral scars or facial nerve palsy) [4]. Loosening of metal plates or screws which might not be visible on radiographs also demand removal [5–7]. According to the literature there are a few reasons where removing of metal plates should be performed: metallosis, corrosion, thermal dysaesthesia, difficulties with future radiological diagnosis, malpositioning, and the migration of osteosynthesis material, particularly in craniofacial surgery [8,9].

Insertion of resorbable screws demand tapping because of the mechanical properties of the material. In long screws there is high risk of fracture of the neck of the screw. Cutting of the threads take much time and any other additional manipulation is linked with the possibility of displacement of condylar fragments. In the "bone welding" method, screws are not inserted such as self-tapping screws but resorbable pins are placed by ultrasound activation [10–13]. The pin is put into the drilled hole and link with osseointegrate with spongy bone.

There are various ways how to perform rigid internal fracture fixation in the condylar area (base, middle, high neck, and head) and it significantly changed in the last decade. To our knowledge, there are no published reports concerning what fixation system and which material provide functionally stable fixation in fractures of the condylar neck of the mandible.

The aim of the study is to compare by means of finite element analysis (FEA) one of the bioresorbable type of "A" shape condyle plate and screws made of polylactic acid (PLLA) versus titanium grade 5 alloy (Ti6Al4V).

2. Materials and Methods

2.1. The Plate

Recently developed resorbable plate (ChM company, Lewickie, Poland, www.chm.eu) has been tested. Manufacturer applied their own new PLLA polymer. Figure 1 shows the design of A-shape condylar plate (ACP) made of resorbable material. The height and the width of the PLLA plate are 31 mm and 19 mm, respectively, whereas the thickness is 1.0 mm. The first bridge of ACP is parallel to posterior border of the mandibular ramus (compression area) and the second is parallel to the sigmoid notch (traction area). The width of the bridge is 2.5 mm. Bridges of ACP are strictly designed according to known compression and traction lines in the mandibular ramus (Figure 2) [14]. There are three linearly located holes on inferior tails of the bridges. Superiorly, the bridges converge and has triangular 3 hole group. Two tails are connected by means of 2.0 mm wide bar. The connecting bar has been designed according to compression line of condylar neck [15].

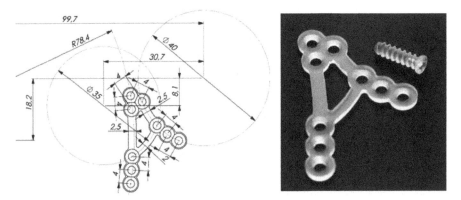

Figure 1. Right side shows the photo of resorbable polylactic acid (PLLA) A-shape right plate and screw whereas on the left side sketch with the dimensions. The height and the width of "A" shape plate are 31 mm and 19 mm respectively whereas the thickness is 1.0 mm. The width of the bridge is 2.5 mm. There are three linearly located holes on inferior tails of the bridges. Superiorly, the bridges converge and has triangular 3-hole group. Two tails are connected by means of 2.0 mm wide bar.

2.2. Material Properties

Mechanical properties have been implemented to calculations are following: Young modulus for Ti-6Al-4V Grade 5 is 140 GPa, Poisson coefficient was 0.3, Yield stress Re = 880 MPa, ultimate tensile stress Rm = 960 MPa, plasticity modulus Ep = 1.4 GPa. For polymer material PLLA: Young modulus 3.5 GPa, Poisson coefficient was 0.35 [16], durability of stretch from 195 MPA to 240 MPa. For mandible material following data were adopted: Young modulus 14 GPa, Poisson coefficient 0.28. Bone is the material that is not described by Hooke's law and it does not have neither linearly elastic characteristic nor plasticity modulus.

Static tensile (Instron 5989 with extensometer, Zwick/Roell, Ulm, Germany) test according to ISO 6892-1 [16] was performed in order to verify the properties of Ti-6Al-4V grade 5. The test results are shown in the graph above (Figure 2). Properties of bone and PLLA have been set on the basis of the literature [17].

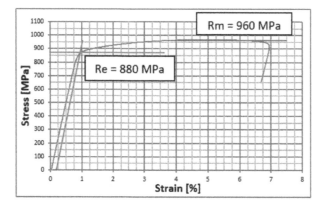

Figure 2. Stress strain curve for the Ti-6Al-4V Grade 5 titanium alloy.

2.3. Finite Element Analysis

Finite element models are helpful for simulating endurance of 3D complex models [18–20].

In the study FEM (Finite Element Method) was used for the comparison between bone end movements just after the open rigid internal fixation by titanium and PLLA plates.

During stretching bone is nonlinearly elastic and after achieving endurance on stretching or squeezing it breaks. In real life bone is not isotropic material but orthotropic, with precise mechanical properties that are different in each layer and maximal load forces in bone depends on direction of loading. Mathematical methods especially for orthotropy are linearly-elastic till the moment of bone destruction. There are several restrictions that in significant way influence on FEM results. Drawing conclusions are made on the basis of the FEM researcher and modelling of biomechanical structure and degree of its simplicity was the basis for the researchers' team. Discussion about FEM results presented as comparison between titanium osteosynthesis and PLLA for the same one.

Authors of this study know advantages and disadvantages of finite element method and restrictions of this method. The software for preparing FEM and simulation was Ansys version 18.2 (Ansys Inc., Canonsburg, PA, USA, www.ansys.com). The proper name of used element is SOLID187 individual name for ANSYS. Users of other software might not know this name whereas tetrahedron tet10 is used for other systems. Mathematical model FEM was solved with material and construction unlinearity using significant displacements. Total size of the quest was DOF (degree of freedom) = 4350480.

2.4. Models

In the Figure 3 mandible 3D model was shown with fixation bone ends in high fracture of condylar neck. Osteosynthesis was designed according to physiology and loaded by forces (purple colour shows forces from muscles) direction and value described by Ramos [21].

In real fractures ragged contours are observed, whereas mathematical model of friction refers to flat surfaces that irregularities are not designed and surface is described as coefficient of friction. In the study to simulate surface degradation by fracture not physical, a (extremely high) coefficient of friction of $\mu = 0.8$ was adopted. The highest coefficient of friction that is gained in machine construction is $\mu = 0.65$, found in brake blocks in aircraft.

The plates were fixed with mandible on basis of continuous displacement. There is initial state of strain between screws and bone because of screw fixation. Contact situation can decrease the stiffness of the connection screw–bone in order to save some time on calculations authors have decided to eliminate the threads on the screw and perform continuous displacements between simplified model of the screw and bone.

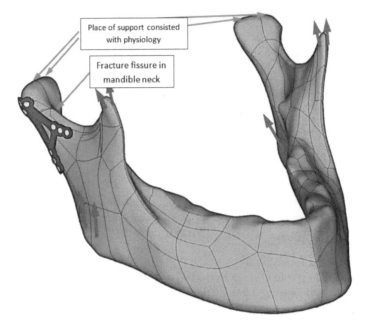

Place of support consisted with physiology

Fracture fissure in mandible neck

Figure 3. 3D model with plate fixation of high neck condylar fracture (red indicators). Fixation has its physiological support points and it was loaded by muscles forces (purple indicators). In the Figure 3 purple indicators have been described in the literature [20].

3. Results

Authors compared the state of bone ends in contact, pressure in contact, displacement in contact, dimension of fissure in contact and comparison maximal equivalent stress distribution in mandible and in osseofixation system.

In the Figure 4a in titanium fixation half of contact area is marked in yellow color. It means that in such a fixation contact is open. Open contact is the reason of so called mechanical silence which might block healing of fracture. However, it is possible that in such conditions collagen mesh may appear and vessels from the region may grow. As a result callus becomes calcified.

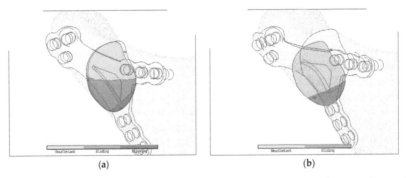

(a) (b)

Figure 4. Contact status just after following plate fixation in the maximal loading according to the literature [16]: (a) Ti-6Al-4V plate fixation—insignificant displacement at the expense of high contact surface of bone ends during loading thanks to which appearance of minimal twisting of bone ends appeared, (b) PLLA plate fixation—significant displacement of bone ends thanks to which high twist appeared.

Orange color shows bone in contact but it is slip so there is possibility of displacement. Cutting forces cause displacement of bone ends and also destroy the gentle collagen mesh preventing calcification, that would never calcified. As a result bone does not heal properly.

Red color describes condition where both parts of bones do not displace to each other. As Figure 5b proves in PLLA fixation is much worse environment that obstruct the healing process in comparison to titanium alloy plate.

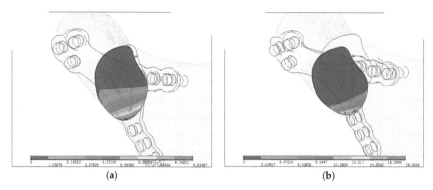

Figure 5. Displacements of pressure in mandibular bone ends just after plate fixation in the maximal loading according to the literature [16]: (**a**) Ti-6Al-4V plate fixation, p_{max} = 9.8 MPa, (**b**) PLLA plate fixation, p_{max} = 18.3 MPa.

As Figure 5a shows surface pressure in contact p_{max} = 9.8 MPa are almost twice lower than in PLLA p_{max} = 18.3 MPa but the area of contact is significantly higher in plate made in Ti-6Al-4V than in PLLA, shown on Figure 5b. Greater area of surface pressure makes healing process faster and as a result it proves osseofixation stiffness.

In Figure 6 displacements of bone ends in fracture plane are shown. Blue area shows the fissure and relative displacements in plane is indeterminable. Displacement on contact surface in Ti-6Al-4V plate is u_{max} = 0.01 mm and is 10 times lower than PLLA plate u_{max} = 0.1 mm.

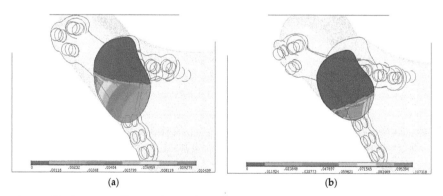

Figure 6. Displacements of bone ends just after plate fixation in the maximal loading according to the literature [16]: (**a**) Ti-6Al-4V plate fixation, u_{max} = 0.01 mm, (**b**) PLLA plate fixation, u_{max} = 0.1 mm.

In Figure 7 size of fissures in plate fixation were compared. Red color means that surface is in contact. Colors from orange to blue means fissure is getting wider. In Ti-6Al-4V plate the widest fissure is g_{max} = 0.0057 mm whereas in PLLA plate fixation it is g_{max} = 0.024 mm.

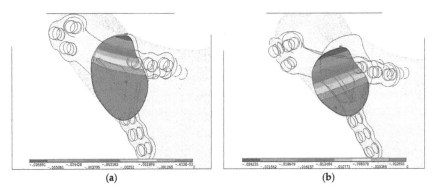

(a) **(b)**

Figure 7. The size of the fissure between bone ends just after plate fixation in the maximal loading according to the literature [16]: (**a**) Ti-6Al-4V plate fixation, g_{max} = 0.0057 mm, (**b**) PLLA plate fixation, g_{max} = 0.024 mm.

Color from blue to red means increasing stresses from 0 to maximum marked on red. In the Figure 8, reduced stresses distribution according to von Mises on PLLA plate with fixing screws made in Ti-6Al-4V and PLLA were the following: $\sigma_{red\,max}$ = 234 MPa and $\sigma_{red\,max}$ = 76 MPa respectively. Reduced stresses are 4 times lower in PLLA screws than in Ti-6Al-4V grade 5 plate. The fact of paramount importance is that Young modulus for Ti-6Al-4V grade 5 is forty times higher than PLLA. Plate made in more flexible material (PLLA) is prone to higher deformation than Ti-6Al-4V. Titanium alloy grade 5 allow for higher loading forces and make the osseofixation much rigid. Thus titanium alloy material make the healing process easier because of the immobilization of the bone ends. In any of discussed situation, yield stress was not exceeded.

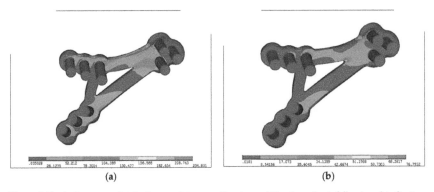

(a) **(b)**

Figure 8. Equivalent stress distribution on plates according to von Mises hypothesis following plate fixation in the maximal loading according to the literature [16]: (**a**) Ti-6Al-4V plate fixation, $\sigma_{red\,max}$ = 234 MPa, (**b**) PLLA plate fixation, $\sigma_{red\,max}$ = 76 MPa; Colors from blue to red means stresses from 0 to maximum.

In the Figure 9, reduced stresses distribution according to von Mises on Ti-6Al-4V and PLLA plate were $\sigma_{red\,max}$ = 87 MPa and PLLA—$\sigma_{red\,max}$ = 91 MPa, respectively. In both cases stresses are comparable. In order to precise the difference of stresses on bone after fixation, strain has been reduced to 30 MPa.

Grey color on Figure 10 shows strain above the scale. Comparing 10(a) and 10(b) figures Ti-6Al-4V plate and screws indicates the center of strains is around last screws in lower part of mandible. In PLLA plate the highest stress is around the hole that is located near the fracture. This location is at risk during surgical fixation because of secondary fractures during plate fixation that restricts possibility of PLLA usage.

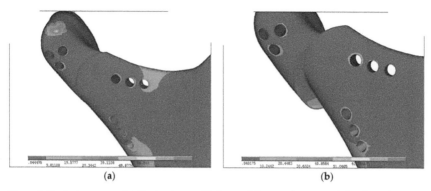

Figure 9. Equivalent stress distribution in mandibular condyle according to von Mises hypothesis just after plate fixation in the maximal loading according to the literature [16]: (**a**) Ti-6Al-4V plate fixation, $\sigma_{red\ max}$ = 87MPa, (**b**) PLLa plate fixation, $\sigma_{red\ max}$ = 91MPa; Colors from blue to red means increasing stresses from 0 to maximum.

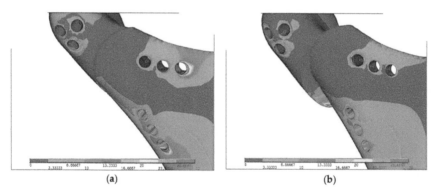

Figure 10. Reduced stress distribution in mandibular condyle according to von Mises hypothesis just after plate fixation in the maximal loading according to the literature [16]: (**a**) Ti-6Al-4V plate fixation, (**b**) PLLA plate fixation; Stresses have been reduced to $\sigma_{red\ max}$ = 30 MPa.

4. Discussion

Bioresorbability is the most desire feature and titanium alloy does not have it. After fixation bone ends osseointegrate in the proper way when the material is useless. Bioresorbable fixation decomposes and the patient's state will be the same as before the surgical procedure "restitution ad integrum". Titanium alloys that are commonly used in osteosynthesis are deprived of such a feature. Implantation of titanium fixation stays with a patient till the end of the life unless it was screwed out by the surgeon. In some cases especially when a patient grows implantation of plate and screws demand removal immediately after bone healing. It required second surgery made in postsurgical scar and facial nerve branch re-preparation (Figure 1). That may lead to permanent facial nerve palsy, thus bioresorbable materials have found its application in maxillofacial surgery.

Surgeons often during the operation in the most cases tighten in the screws without dynamometer key. It might be assumed that every single screw might be screwed with different torque and as a result stress on every screw is different. This matter has been omitted in order to simplify the procedure as well. Authors have focused on modelling of the contact of the bone ends just after the fracture and comparison of fixation materials taking into consideration only contact between bone ends. On the basis of computed tomography 3D model of edentulous mandible has been designed. Loading on the model is described as loading area transformed to the notches. Supports in temporomandibular joint have been defined as superior notch and linked on the real contact surface by Rigid Body Element (RBE). Such modelling ensures free rotation around joint. On the surface where earlier geometrically teeth have been extracted, degree of freedom are linked with the process of biting (in the direction perpendicular to the surface where teeth have been extracted).

Just after fixation, a patient who fulfils diet precautions and low forces on chewing is able to function almost normally. However, if a surgeon wants to achieve total biodegradability of fixation, then PLLA is the material of choice. The most important factor regarding durability is to perform stresses less than yield stress in order to prevent from irreversible deformation of screws and plates. That is why maxillofacial surgeon asks patients not to overload jaws regarding biting tough food especially after the surgery. Process of making new bone takes about 2 weeks and after this time new bone starts to load forces [22]. By using PLLA materials time of compulsory low forces chewing lengthens from a few days to a few months.

Plates made in PLLA or PLGA are biodegradable in comparison to Ti-6Al-4V. In terms of pressure in fracture, displacement, strain condition Ti-6Al-4V is superior to polymer/composite materials apart from biodegradability. PLLA osteosynthesis indicates that in comparison to titanium alloy fixation contact surface plate with slightly sliding between bones and plates is only 30%. Moreover, displacement between bones has occurred. In the Figure 3 it is easy to notice that in titanium fixation, so called mechanical silence occurs on half of the area after fracture whereas in case of PLLA plate only 20% of the area is in mechanical silence thus bone healing is not possible. Comparing PLLA to titanium alloy small area of concentration of the forces were observed. That is why pressure on bone in PLLA osteosynthesis is twice higher than in titanium fixation. On small areas where pressure on bone is too high it may cause local bone degradation around the fracture (i.e., bone atrophy by compression) and it might be the reason of delay in the healing process or make it impossible. The same conclusion might be assumed by observing cutting forces in the fracture shown in the Figure 4. The higher forces the higher displacement in fracture plane. In significant displacement of bone fragments in fracture fissure, collagen fibres tear and inhibit the healing process (Figure 5). Using PLLA plate has another negative feature if it comes to displacement in fracture plane. Osteosynthesis by means of PLLA is such flexible that strong biting might cause opening of fracture fissure resulting in sliding and twisting that lead to degradation of callus. Using PLLA plates require additional immobilization such as intermaxillary traction or liquid diet. Fixation cannot be overloaded otherwise bone healing process was inhibited.

Author Contributions: Conceptualization, M.K. and J.Ś.; methodology, J.Ś.; software, J.Ś.; validation, J.Ś., M.K. and R.Z.; formal analysis, M.K.; investigation, R.Z.; resources, J.Ś.; data curation, J.Ś.; writing—original draft preparation, R.Z.; writing—review and editing, R.Z., J.Ś., M.K.; visualization, J.Ś., R.Z., M.K.; supervision, J.Ś., M.K.; project administration, M.K., J.Ś.; funding acquisition, M.K.

Funding: This research was funded by Medical University, grant number 503/5-061-02/503-51-001.

Acknowledgments: ChM (www.chm.eu) for support the study.

Conflicts of Interest: The authors declare no conflict of interest.

References

1. Eckelt, U.; Schneider, M. Open versus closed reduction of condylar process fractures—A prospective randomized multicenter study with regard to fracture localisation. *Int. J. Oral Maxillofac. Surg.* **2009**, *38*, 478. [CrossRef]
2. Eckelt, U.; Schneider, M.; Erasmus, F.; Gerlach, K.L.; Kuhlisch, E.; Loukota, R.; Rasse, M.; Schubert, J.; Terheyden, H. Open versus closed treatment of fractures of the mandibular condylar process—A prospective randomized multi-center study. *J. Craniomaxillofac. Surg.* **2006**, *34*, 306–314. [CrossRef]

3. Hlawitschka, M.; Loukota, R.; Eckelt, U. Functional and radiological results of open and closed treatment of intracapsular (diacapitular) condylar fractures of the mandible. *Int. J. Oral Maxillofac. Surg.* **2005**, *34*, 597–604. [CrossRef] [PubMed]

4. Eckelt, U.; Hlawitschka, M. Clinical and radiological evaluation following surgical treatment of condylar neck fractures with lag screws. *J. Craniomaxillofac. Surg.* **1999**, *27*, 235–242. [CrossRef]

5. Schneider, M.; Lauer, G.; Eckelt, U. Surgical treatment of fractures of the mandibular condyle: A comparison of long-term results following different approaches—Functional, axiographical, and radiological findings. *J. Craniomaxillofac. Surg.* **2007**, *35*, 151–160. [CrossRef] [PubMed]

6. Klotch, D.W.; Lundy, L.B. Condylar neck fractures of the mandible. *Otolaryngol. Clin. N. Am.* **1991**, *24*, 181–194.

7. Choi, B.H.; Yoo, J.H. Open reduction of condylar neck fractures with exposure of the facial nerve. *Oral Surg. Oral Med. Oral Pathol. Oral Radiol. Endod.* **1999**, *88*, 292–296. [CrossRef]

8. Lizuka, T.; Lindqvist, C. Rigid internal fixation of mandibular fractures. An analysis of 270 fractures treated using the AO/ASIF method. *Int. J. Oral Maxillofac. Surg.* **1992**, *21*, 65–69.

9. Kim, Y.K.; Yeo, H.H.; Lim, S.C. Tissue response to titanium plates: A transmitted electron microscopic study. *J. Oral Maxillofac. Surg.* **1997**, *55*, 322–326. [CrossRef]

10. Pilling, E.; Mai, R.; Theissig, F.; Stadlinger, B.; Loukota, R.; Eckelt, U. An experimental in vivo analysis of the resorption to ultrasound activated pins (Sonic weld) and standard biodegradable screws (ResorbX) in sheep. *Br. J. Oral Maxillofac. Surg.* **2007**, *45*, 447–450. [CrossRef] [PubMed]

11. Eckelt, U.; Nitsche, M.; Muller, A.; Pilling, E.; Pinzer, T.; Roesner, D. Ultrasound aided pin fixation of biodegradable osteosynthetic materials in cranioplasty for infants with craniosynostosis. *J. Craniomaxillofac. Surg.* **2007**, *35*, 218–221. [CrossRef]

12. Meissner, H.; Pilling, E.; Richter, G.; Koch, R.; Eckelt, U.; Reitemeier, B. Experimental investigations for mechanical joint strength following ultrasonically welded pin osteosynthesis. *J. Mater. Sci. Mater. Med.* **2008**, *19*, 2255–2259. [CrossRef] [PubMed]

13. Pilling, E.; Meissner, H.; Jung, R.; Koch, R.; Loukota, R.; Mai, R.; Reitemeier, B.; Richter, G.; Stadlinger, B.; Stelnicki, E.; et al. An experimental study of the biomechanical stability of ultrasound-activated pinned (SonicWeld Rx + Resorb-X) and screwed fixed (Resorb-X) resorbable materials for osteosynthesis in the treatment of simulated craniosynostosis in sheep. *Br. J. Oral Maxillofac. Surg.* **2007**, *45*, 451–456. [CrossRef]

14. Meyer, C.; Kahn, J.L.; Boutemi, P.; Wilk, A. Photoelastic analysis of bone deformation in the region of the mandibular condyle during mastication. *J. Cranio-Maxillofac. Surg.* **2002**, *30*, 160–169. [CrossRef]

15. Kozakiewicz, M.; Swiniarski, J. "A" shape plate for open rigid internal fixation of mandible condyle neck fracture. *J. Cranio-Maxillofac. Surg.* **2014**, *42*, 730–737. [CrossRef]

16. *Metallic Materials—Tensile Testing—Part 1: Method of Test at Room Temperature*; ISO 6892-1; International Organization for Standardization: Geneva, Switzerland, 2016.

17. Maurer, P.; Holweg, S.; Knoll, W.D.; Schubert, J. Study by finite element method of the mechanical stress of selected biodegradable osteosynthesis screws in sagittal ramus osteotomy. *Br. J. Oral Maxillofac. Surg.* **2002**, *40*, 76–83. [CrossRef] [PubMed]

18. Hart, T.R.; Hennebel, V.; Thongpreda, N.; Van Buskirk, W.C.; Anderson, R.C. Modelling the biomechanics of the n audible: A three-dimensional finite element study. *J. Biomech.* **1992**, *25*, 261–286. [CrossRef]

19. Field, C.; Ichim, I.; Swain, M.V.; Chan, E.; Darendeliler, M.A.; Li, W.; Li, Q. Mechanical responses to orthodontic loading: A 3-dimensional finite element multi-tooth model. *Am. J. Orthod. Dentofacial. Orthop.* **2009**, *135*, 174–181. [CrossRef] [PubMed]

20. Ichim, I.; Kieser, J.A.; Swain, M.V. Functional significance of strain distribution in the human mandible under masticatory load: Numerical predictions. *Arch. Oral Biol.* **2007**, *52*, 465–473. [CrossRef] [PubMed]

21. Ramos, A.; Completo, A.; Relvas, C.; Mesnard, M.; Simoes, J.A. Straight, Semi-anatomic and anatomic TMJ implants: The influence of condylar geometry and bone fixation screws. *J. CranMaxillofac. Surg.* **2011**, *39*, 343–350. [CrossRef]

22. Marsell, R.; Einhorn, T. The biology of fracture healing. *Injury* **2011**, *42*, 551–555. [CrossRef] [PubMed]

Article

Does the Hirsch Index Improve Research Quality in the Field of Biomaterials? A New Perspective in the Biomedical Research Field

Saverio Affatato * and Massimiliano Merola

Laboratorio di Tecnologia Medica, IRCCS-Istituto Ortopedico Rizzoli, Via di Barbiano, 1/10 40136 Bologna, Italy; massimiliano.merola@tecno.ior.it
* Correspondence: affatato@tecno.ior.it; Tel.: +39-051-6366864; Fax: +39-051-6366863

Received: 21 September 2018; Accepted: 11 October 2018; Published: 13 October 2018

Abstract: Orthopaedic implants offer valuable solutions to many pathologies of bones and joints. The research in this field is driven by the aim of realizing durable and biocompatible devices; therefore, great effort is spent on material analysis and characterization. As a demonstration of the importance assumed by tribology in material devices, wear and friction are two of the main topics of investigation for joint prostheses. Research is led and supported by public institutions, whether universities or research centers, based on the laboratories' outputs. Performance criteria assessing an author's impact on research contribute somewhat to author inflation per publication. The need to measure the research activity of an institution is an essential goal and this leads to the development of indicators capable of giving a rating to the publication that disseminates them. The main purpose of this work was to observe the variation of the Hirsch Index (h-index) when the position of the authors is considered. To this end, we conducted an analysis evaluating the h-index by excluding the intermediate positions. We found that the higher the h value, the larger the divergence between this value and the corrected one. The correction relies on excluding publications for which the author does not have a relevant position. We propose considering the authorship order in a publication in order to obtain more information on the impact that authors have on their research field. We suggest giving the users of researcher registers (e.g., Scopus, Google Scholar) the possibility to exclude from the h-index evaluation the objects of research where the scientist has a marginal position.

Keywords: h-index; bibliometric indicators; biomaterials; quality of research; citations

1. Introduction

Joint replacement surgery is a successful and consolidated branch of orthopaedics. Its progressive achievement in alleviating pain and disability, helping patients to return to an active life, is reliant on efficient relationships between clinicians and researchers working across transverse areas of medicine and science [1]. The purpose of tribology research applied to orthopaedics is the minimization and elimination of losses resulting from friction and wear [2]. The research of new biomaterials plays an important role, and as a consequence, in vitro tests for such materials are of great importance [3]. The knowledge of the laboratory wear rate is an important aspect in the preclinical validation of prostheses. Research and development of wear-resistant materials continues to be a high priority [4–6]. Clinical research designed to carefully evaluate the performance of new materials intended to reduce wear is essential to ascertaining their efficacy and preventing the possibility of unexpected failure [7,8]. Unfortunately, failures and revision surgeries still constitute the main clinical problems related to total joint replacement [9]. The research is therefore constantly pushed to find new solutions to wear-related issues and to identify new high-standard materials. The public institutions of research receive national funding on the bases of their results and are therefore constrained to obtain high levels of quality assessment [10]. Evaluation of scientific publications is the criterion used by the universities and research institutes to measure the merit and value of researchers and academics [11], and it has a crucial impact on research funds distribution [12–14]. The need to measure the research quality of institutions is an essential goal, which has led to the development of indicators capable of giving a rating to publications. These indicators are used in bibliometric disciplines to quantitatively evaluate the quality and diffusion of scientific production within the scientific community. In order to obtain funds within the orthopaedic community, there is a strong pressure on researchers to publish even if the merit of the study is unreliable; this is because objectives are set to achieve a certain number of publications instead of focusing on the quality of the research [15].

There are two main ways to evaluate scientific research:

- Bibliometric indicators are quantitative methods based on the number of times a publication is cited. The higher the number of citations, the larger the group of researchers who have used this work as a reference and, thus, the stronger its impact on the scientific community;
- Peer review is a qualitative method based on the judgement of experts. A small number of researchers, specialized in the field of the work, analyze and evaluate the scientific value of a publication.

Eugene Garfield proposed the Impact Factor (IF) in 1955 [16,17] with the intent to help scientists look for bibliographic references; the IF indicator was quickly adopted to assess the influence of journals and, not long after, of individual scientists. A journal's impact factor is the ratio of two elements: the numerator is the number of citations in the current year to items published in the previous two years, and the denominator is the total number of articles published in the same two years [16,17]. It is published by the ISI-Thomson publisher on the basis of the Web of Science database and measures the frequency with which an article published in a journal is cited by other periodicals over a specific period of time (two years after its release). This measure is used as an appraisal of the importance of a magazine compared with others in the same sector: the higher the impact factor, the more authoritative the magazine [18]. The impact factors of each magazine can be consulted on the website of the Journal of Citation Reports (JCR) [19].

The impact factor is widely used as an index of academic research quality. It is also applied as a winning criterion for the granting of funds and incentives, or as a basis for the evaluation of a scholar or a professional in public competitions [11]. Yet the impact factor is an indicator for measuring the impact of a magazine in its specific disciplinary area, certainly not for evaluating the authors, and this latter use has been criticized for many reasons. In order to overcome the problems of the impact factor, in 2005, Jorge E. Hirsch [20] proposed a new index, known as the h-index or Hirsch index, as a single-number criterion to evaluate the scientific output of a researcher. It combines a measure of quantity (publications) and impact (citations). In other words, a scientist has index h if h-many of his articles have at least h-many citations each, and the other articles have fewer than h-many citations each [21]. It performs better than other single-number criteria previously used to evaluate the scientific output of a researcher (impact factor, total number of documents, total number of citations, citation per paper rate, and number of highly cited papers) [22]. The h-index is easy to understand and can be easily gained by anyone with access to the Thomson ISI Web of Science [22]. Actually, the h-index has been reviewed as one of the most reliable criteria for evaluating the scientific outputs of researchers [23]. This index has many flaws: the articles with citations less than the h-index value are excluded from the calculation. The number of citations is influenced by self-citations and colleague citations, meaning that its value can be increased by recommendation to friends and colleagues [23–25]. There are many circumstances where the h-index provides misleading information on the impact of an author. However, this popular bibliometric indicator does not consider multiple co-authorship nor author position [11,26]. To our knowledge, no policy guides author order in biomedical publications. The position of an author, controversies about author order, and disagreements on the involvement of the last author are constantly debated; thus, it is worth analyzing the relationship between author position and bibliometric indicators [27]. The contribution of an author to a research project is not always clear, especially when a manuscript is attributed to a large group [28].

Rating and weighting of research value is not an easy task, as history shows; often, what is believed to be modern and mainstream research will yield highly rated papers but is not always a guarantee of innovation and scientific progress [15]. With this in mind, and to go more into depth in this matter, we produced a statistical evaluation of the h-index on a cohort of 60 authors belonging to the biomedical field, accounting for only given positions. In detail, three modified h-values were evaluated: considering only the First (F) and Second (S) authorships (referred to as FS); only the F and the Last (L) authorships (referred to as FL); and the F, S, and L positions (referred to as FSL). The main goal of this work was to implement an algorithm that can calculate a modified h-index considering the author's position in an article.

Different approaches are available to decide the author order, like sorting them alphabetically or listing them in descending order according to their contribution [29]. Several approaches may be used to assess the contribution of an author to a paper [30]. In the "sequence-determines-credit" (SDC) system, the author order reflects the declining importance of their contribution. The "equal contribution" approach uses alphabetical order and implies identical involvement. The "first-last-emphasis" norm underlines the importance of the last author. Using the "percent-contribution-indicated" implies detailing each author's impact.

The convention used in the biomedical field of research, as reported in the literature [27], is as follows: the first author conducts the majority of the work; the last author could be the senior member of the group and usually leads the research; co-authors, those between the first and last, are ranked in order of their input to the work; the corresponding authors—typically senior scholars—communicate with editors and readers. With this premise, we considered the first, the second, and the last authors as the main contributors to biomedical research.

2. Methods

A total of 60 authors from the biomedical field were selected as a cohort; 30 of them were obtained from an Italian scientist ranking system [31] and the other 30 were chosen from all around the world.

To obtain the modified h-index, an algorithm was implemented in MATLAB (MathWorks, Natick, MA, USA). The complete list of articles attributed to an author was extracted from the scopus.com webpage using the implemented feature called "Export all". The information extracted was limited to Authors, Title, and Citation Count. Each list was obtained in the csv extension; these lists were then imported into the workspace of MATLAB using the Import Tool App. Each author entry was analyzed to find out their position; thus, according to the excluding criterion, the publication was considered or not for the evaluation of h. Given that authors are frequently cited in different ways (e.g., Surname, Name; Surname, N.; Surname M.N., etc.), the investigation considered all different options. If a publication respected the inclusion criterion, its citation count was taken into account. The list of included citation counts was sorted in ascending order; thus, the h value was by evaluated considering the last position at which the citation count was higher than or equal to the position itself. In Figure 1, a flow chart of the exclusion process is shown.

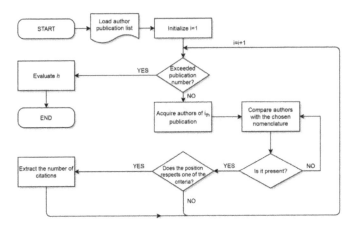

Figure 1. Flowchart of the process to obtain the corrected h-value.

3. Results

In Figure 2 are summarized the results obtained through the different exclusion criteria here studied. It is worth underlining the large divergences obtained comparing the higher h-values with their respective corrected ones.

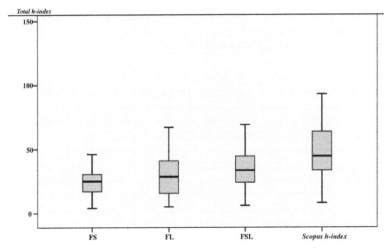

Figure 2. A box plot shows the modified h-index values (± standard deviation) for all authors considered in this study. The total h-index retrieved from Scopus is the highest of the four classifications.

From the results in Figure 2, we emphasize that on the basis of our exclusion criteria, the modified h-value decreases significantly.

To highlight these differences, the entire cohort was divided into three sub-cohorts. A first group of authors with h-value ranging from 0 to 35 (called Low), a second group from 36 to 50 (Middle), and a third group from 51 up to the maximum of 181 (High) were extracted, as shown in Figure 3.

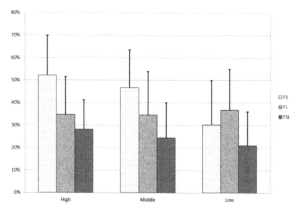

Figure 3. Histogram of the influence of the exclusion criteria on the h-values in sub-cohorts based on the starting h from Scopus.

In Figure 3 it is underlined that the High sub-cohort, starting from the highest value of h, is the most affected by all the exclusion criteria. This is especially true in the FS case, where it reaches a mean reduction of more than 50%. On the contrary, for the Low group, this exclusion criterion only decreases the h-value by roughly 30%. The Low group is more influenced by the exclusion of the publications where the authors are present as second authors (35% of decrease). The Middle group is more affected by the FS criterion, followed by the FL and the FSL.

In Figure 4, another histogram of the influence of the exclusion criteria is presented. In this case, the sub-cohorts were obtained based on the number of articles each author has. Considering the large range (from a minimum of 15 to a maximum of 1241), we chose to obtain a further group, yielding a total of four. The Low group collected the authors with up to 150 publications, the Middle group ranged from 151 to 300, High from 301 to 500, and Super High (S. High) from 501 to the maximum.

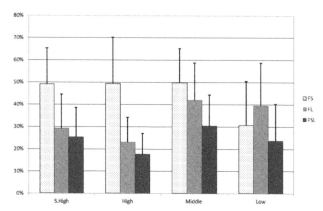

Figure 4. Histogram of the influence of the exclusion criteria on the h-values with sub-cohorts based on the number of publications.

This representation also outlined how the authors with a great number of publications are more affected by the exclusion criteria that do not take into consideration the articles where they are listed last among the authors. The first three sub-cohorts have a mean reduction of around 50% from their starting h-value when the FS criterion is applied. On the counter side, the FL criterion more greatly affects the Middle and the Low cohorts, reaching roughly 40% against the 30% and 20% of the Super

High and High, respectively. The FSL criterion has a similar outcome on the group, but is stronger on the Middle and the Super High sub-cohorts, where it is about 30%.

4. Discussion and Conclusions

The term "impact factor" has gradually evolved to describe both journals' and authors' impacts. Journal impact factors generally involve relatively large populations of articles and citations. There are some ambiguities in the use of the h-index that could provoke changes in the publishing behaviour of scientists, such as increasing the number of self-citations distributed among the documents on the edge of the h-index [32]. Another disadvantage of the h-index is that not all citations of a researcher are involved in the calculation and its value may not increase with a rise in citations. Highly cited papers are significant for the evaluation of the h-index, but once they belong to the top h-many papers, the effect of number of citations they get is negligible [22]. Considering that the evaluators reduce scientific success to a single value, researchers can change their behaviour to increase these values, even by using unethical strategies [33]. Moreover, the scientific independent criterion of peer-review evaluation is going to be replaced by a system of private companies whose only feedback is the indicator number [11]. The authors believe that it is essential to continue analyzing the bibliometric indicators in order to establish their drawbacks and limitations and to propose improvements where necessary. It is especially relevant to determine in which cases this index could be biased, since it could have serious consequences for the assessment of scientists and academics. Our work proposes to be an insight into the bibliometric indicators and to help the scientific community and research institutions consider how authorship position impacts on the h-value.

The main purpose of this work was to observe the variation of the h-value when the position of authorship is considered. Therefore, an empirical analysis was conducted to assess the influence of an author's intermediate positions on their h-index. The results of this study showed that when the h-value is high, there is a large divergence between this value and the "corrected" one. We propose an improved method to consider the importance of the authorship in a publication in order to obtain a more profound understanding of the effective impact that an author has on their research field. This could be realized by giving users of researcher registers (e.g., Scopus, Google Scholar, etc.) the possibility to select which version of the h-index they want to analyze. The users should have the possibility, along with the already-implemented exclusion of self-citations, to exclude the publications where the authors have an intermediate position. It is worth noting that this correction should be considered for senior researchers, whereas young ones often have difficulties obtaining a relevant position in authorship. Thus, this exclusion can negatively affect the process of researcher renewal, discouraging young ones who would not see their efforts rewarded.

We believe that bibliometric indices should evolve with the aim of fairer evaluation of scientific productions. This would also be beneficial for funding distribution in the biomedical and orthopaedic research spheres, above all considering the threat of research cuts being to the detriment of patients' health. Alternatively, there is the risk that research toward new and more efficient biomaterials could stagnate due to a lack of capital granted to deserving scientists.

Author Contributions: S.A. conceived and designed the experiments; M.M. performed the software analyses; S.A. wrote the first draft of the manuscript; S.A. and M.M. analyzed the data; S.A. and M.M. wrote the final paper.

Funding: This research received no external funding.

Acknowledgments: The authors thank Barbara Bordini for her help with statistical analyses.

Conflicts of Interest: The authors declare no conflict of interest.

References

1. Affatato, S. *Perspectives in Total Hip Arthroplasty: Advances in Biomaterials and their Tribological Interactions*; Affatato, S., Ed.; Elsevier Science: New York, NY, USA, 2014.
2. Viceconti, M.; Affatato, S.; Baleani, M.; Bordini, B.; Cristofolini, L.; Taddei, F. Pre-clinical validation of joint prostheses: A systematic approach. *J. Mech. Behav. Biomed. Mater.* **2009**, *2*, 120–127. [CrossRef] [PubMed]
3. Matsoukas, G.; Willing, R.; Kim, I.Y. Total Hip Wear Assessment: A Comparison Between Computational and In Vitro Wear Assessment Techniques Using ISO 14242 Loading and Kinematics. *J. Biomech. Eng.* **2009**, *131*, 41011. [CrossRef] [PubMed]
4. Oral, E.; Neils, A.; Muratoglu, O.K. High vitamin E content, impact resistant UHMWPE blend without loss of wear resistance. *J. Biomed. Mater. Res. B Appl. Biomater.* **2015**, *103*, 790–797. [CrossRef] [PubMed]
5. Ansari, F.; Ries, M.D.; Pruitt, L. Effect of processing, sterilization and crosslinking on UHMWPE fatigue fracture and fatigue wear mechanisms in joint arthroplasty. *J. Mech. Behav. Biomed. Mater.* **2016**, *53*, 329–340. [CrossRef] [PubMed]
6. Kyomoto, M.; Moro, T.; Yamane, S.; Saiga, K.; Watanabe, K.; Tanaka, S.; Ishihara, K. High fatigue and wear resistance of phospholipid polymer grafted cross-linked polyethylene with anti-oxidant reagent. In Proceedings of the 10th World Biomaterials Congress, Montréal, QC, Canada, 17–22 May 2016.
7. Essner, A.; Schmidig, G.; Wang, A. The clinical relevance of hip joint simulator testing: In vitro and in vivo comparisons. *Wear* **2005**, *259*, 882–886. [CrossRef]
8. Affatato, S.; Spinelli, M.; Zavalloni, M.; Mazzega-Fabbro, C.; Viceconti, M. Tribology and total hip joint replacement: Current concepts in mechanical simulation. *Med. Eng. Phys.* **2008**, *30*, 1305–1317. [CrossRef] [PubMed]
9. Ulrich, S.D.; Seyler, T.M.; Bennett, D.; Delanois, R.E.; Saleh, K.J.; Thongtrangan, I.; Stiehl, J.B. Total hip arthroplasties: What are the reasons for revision? *Int. Orthop.* **2008**, *32*, 597–604. [CrossRef] [PubMed]
10. Geuna, A.; Martin, B.R. University Research Evaluation and Funding: An International Comparison. *Minerva* **2003**, *41*, 277–304. [CrossRef]
11. Carpenter, C.R.; Cone, D.C.; Sarli, C.C. Using publication metrics to highlight academic productivity and research impact. *Acad. Emerg. Med.* **2014**, *21*, 1160–1172. [CrossRef] [PubMed]
12. Rezek, I.; McDonald, R.J.; Kallmes, D.F. Is the h-index Predictive of Greater NIH Funding Success Among Academic Radiologists? *Acad. Radiol.* **2011**, *18*, 1337–1340. [CrossRef]
13. Cirimina, R.; Pagliaro, M. On the use of the h-index in evaluating chemical research. *Chem. Central J.* **2013**, *7*, 132. [CrossRef] [PubMed]
14. Narin, F.; Olivastro, D.; Stevens, K.A. Bibliometrics: Theory, practice and problems. *Eval. Rev.* **1994**, *18*, 65–76. [CrossRef]
15. Fayaz, H.C.; Haas, N.; Kellam, J.; Bavonratanavech, S.; Parvizi, J.; Dyer, G.; Smith, M. Improvement of research quality in the fields of orthopaedics and trauma—A global perspective. *Int. Orthop.* **2013**, *37*, 12051212. [CrossRef] [PubMed]
16. Garfield, E. The history and meaning of the journal impact factor. *JAMA* **2006**, *295*, 90–93. [CrossRef] [PubMed]
17. Garfield, E. The meaning of the Impact Factor. *Int. J. Clin. Health Psychol.* **2003**, *3*, 363–369.
18. Bordons, M.; Fernández, M.T.; Gómez, I. Advantages and limitations in the use of impact factor measures for the assessment of research performance. *Scientometrics* **2002**, *53*, 195–206. [CrossRef]
19. How Do I Find the Impact Factor and Rank for a Journal?. Available online: https://guides.hsl.virginia.edu/faq-jcr (accessed on 12 October 2018).
20. Hirsch, J.E. An index to quantify an individual's scientific research output. *Proc. Natl. Acad. Sci. USA* **2005**, *102*, 16569–16572. [CrossRef] [PubMed]
21. Bornmann, L.; Daniel, H.D. Does the h-index for ranking of scientists really work? *Scientometrics* **2005**, *65*, 391–392. [CrossRef]
22. Costas, R.; Bordons, M. The h-index: Advantages, limitations and its relation with other bibliometric indicators at the micro level. *J. Informetr.* **2007**, *1*, 193–203. [CrossRef]
23. Ahangar, H.G.; Siamian, H.; Yaminfirooz, M. Evaluation of the scientific outputs of researchers with similar h index: A critical approach. *Acta Inform. Med.* **2014**, *22*, 255–258. [CrossRef] [PubMed]

24. Martin, B.R. Whither research integrity? Plagiarism, self-plagiarism and coercive citation in an age of research assessment. *Res. Policy* **2013**, *42*, 1005–1014. [CrossRef]

25. Foo, J.Y.A. Impact of excessive journal self-citations: A case study on the folia phoniatrica et logopaedica journal. *Sci. Eng. Ethics* **2011**, *17*, 65–73. [CrossRef] [PubMed]

26. Kreiman, G.; Maunsell, J.H.R. Nine Criteria for a Measure of Scientific Output. *Front. Comput. Neurosci.* **2011**, *5*, 1–6. [CrossRef] [PubMed]

27. Du, J.; Tang, X.L. Perceptions of author order versus contribution among researchers with different professional ranks and the potential of harmonic counts for encouraging ethical co-authorship practices. *Scientometrics* **2013**, *96*, 277–295.

28. Tarkang, E.E.; Kweku, M.; Zotor, F.B. Publication practices and responsible authorship: A review article. *J. Public Health Afr.* **2017**, *8*, 36–42. [CrossRef] [PubMed]

29. Kissan, J.; Laband, D.N.; Patil, V. Author order and research quality. *South. Econ. J.* **2005**, *7*, 545–555.

30. Tscharntke, T.; Hochberg, M.E.; Rand, T.A.; Resh, V.H.; Krauss, J. Author sequence and credit for contributions in multiauthored publications. *PLoS Biol.* **2007**, *5*, 18. [CrossRef] [PubMed]

31. Degli Esposti, M.; Boscolo, L. Top Italian Scientists Biomedical Sciences. 2018. Available online: http://www.topitalianscientists.org/TIS_HTML/Top_Italian_Scientists_Biomedical_Sciences.htm (accessed on 17 April 2018).

32. Van Raan, A.F. Comparison of the Hirsch-index with standard bibliometric indicators and with peer judgment for 147 chemistry research groups. *Scientometrics* **2006**, *67*, 491–502. [CrossRef]

33. Masic, I. H-index and how to improve it? *Donald Sch. J. Ultrasound. Obstet. Gynecol.* **2016**, *10*, 83–89. [CrossRef]

 *materials*

Article

Development of a Novel in Silico Model to Investigate the Influence of Radial Clearance on the Acetabular Cup Contact Pressure in Hip Implants

Saverio Affatato [1],*, Massimiliano Merola [1] and Alessandro Ruggiero [2]

[1]   Laboratorio di Tecnologia Medica, IRCCS—Istituto Ortopedico Rizzoli, 40136 Bologna, Italy; massimiliano.merola@ior.it

[2]   Department of Industrial Engineering, University of Salerno, 84084 Salerno, Italy; ruggiero@unisa.it

*   Correspondence: affatato@tecno.ior.it; Tel.: +39-051-6366-864; Fax: +39-051-6366-863

Received: 18 June 2018; Accepted: 20 July 2018; Published: 25 July 2018

Abstract: A hip joint replacement is considered one of the most successful orthopedic surgical procedures although it involves challenges that must be overcome. The patient group undergoing total hip arthroplasty now includes younger and more active patients who require a broad range of motion and a longer service lifetime of the implant. The current replacement joint results are not fully satisfactory for these patients' demands. As particle release is one of the main issues, pre-clinical experimental wear testing of total hip replacement components is an invaluable tool for evaluating new implant designs and materials. The aim of the study was to investigate the cup tensional state by varying the clearance between head and cup. For doing this we use a novel hard-on-soft finite element model with kinematic and dynamic conditions calculated from a musculoskeletal multibody model during the gait. Four different usual radial clearances were considered, ranging from 0 to 0.5 mm. The results showed that radial clearance plays a key role in acetabular cup stress-strain during the gait, showing from the 0 value to the highest, 0.5, a difference of 44% and 35% in terms of maximum pressure and deformation, respectively. Moreover, the presented model could be usefully exploited for complete elastohydrodynamic synovial lubrication modelling of the joint, with the aim of moving towards an increasingly realistic total hip arthroplasty in silico wear assessment accounting for differences in radial clearances.

Keywords: finite element analysis; total hip arthroplasty; UHMWPE; musculoskeletal multibody model; tribology

---

1. Introduction

Total hip replacement (THR) is the most successful application of biomaterials in the short term in order to alleviate pain, restore joint architecture, and increase functional mobility in diseased traumatized joints [1]. A major limiting factor to the service life of THRs remains the wear of the polyethylene acetabular cup. Preclinical endurance testing has become a standard procedure to predict the mechanical performance of new devices during implant development. Wear tests are performed on materials and designs used in prosthetic implants [1–3] to obtain quality control and acquire further knowledge on the tribological behavior in joint prostheses. To gain realistic results, a wear test should replicate the in vivo working conditions of the artificial implants [4]. However, it is well known that wear tests have a long duration and are expensive [1,5,6]. The simulation is run for several million cycles; assuming that one million cycles correspond to one year in vivo [7–9]. The running-in period encompasses approximately the first half a million cycles [10]; therefore, the steady-state wear is assessed by measuring the wear when the running-in period is over. Wear is measured either gravimetrically, measuring the weight loss of the component, or through a direct analysis of the volume that has been removed, optically or by coordinate-measuring [4,11–13].

Although in vitro wear evaluation of new medical devices is standardized [14], wear tests are not flexible enough. The prediction of wear in hip replacements has been a subject of intense study in recent years [15]. In silico investigation, where algorithms could be developed to model a biomedical process, is a logical extension of controlled in vitro experimentation. It is the natural result of the explosive increase in computing power available to scientists; thus, numerical models could be used to predict results of a wear test with less time and cost. Obviously, the in silico analysis integrates but does not replace the experimental tools.

Finite element analysis (FEA) has been widely used in many areas of biomechanics and biomechanical engineering to study parameters and boundary conditions, which are not accessible experimentally [15,16]. The numerical modeling tool of FEA has been widely applied to analyze the behavior of articular cartilage, joints, and bone structures under compressive and tensile stresses [17–21]. It was also used to improve the design of hip joint prostheses and to minimize experimental tests [22]. Structural applications include the design and development of joint prostheses and fracture fixation devices [23]. To make computational wear simulation meaningful, better wear models are needed, and those can only be built on experimental data that call for systematic experiments in which the relationships between the wear and the material properties, movements, lubrication, and loading are investigated.

FEA is a useful way to investigate the effects of changing parameters such as load, velocity, contact geometry, material properties, etc. This technique was introduced to orthopedic biomechanics in 1972 to evaluate stresses in a human bone [24]. Since then, it has been utilized with increasing frequency in the biomedical engineering context. FEA uses algorithms in which a domain is realized by a number of sub-domains referred to as elements [25,26]. The behavior of each element is readily defined and understood by numerical equations, which allow the study of complex behaviors of the entire body.

Many researchers use FEA to investigate the mechanical and tribological behavior of the most widespread hip implants. Maxian et al. [27] developed an adaptive re-meshing model to study the wear evolution on a long-term regime (20 years of follow up). He found that sliding distance was mainly responsible for the volumetric wear more than polyethylene thickness. Hu and collaborators [28] developed, in 2001, a fully thermo-mechanical coupled finite element model of a total hip prosthesis. The model simulating the wear test in a hip simulator was used to evaluate the transient contact stresses and to predict the rise of temperature due to the friction for different applied loads, sliding speeds, and frictional coefficients. Gao and co-workers [15] developed a wear model, on metal-on-metal hip replacements, which considered lubrication for the first time via a transient elastohydrodynamic lubrication regime. Sfantos et al. [29], on the other hand, performed a parametric study of the wear in total hip arthroplasty by using a boundary elements method.

The aim of this study was to develop a flexible finite element model, starting from the dynamic data retrieved by means of a multibody musculoskeletal model of the lower limb. In particular, in a previous work Ruggiero et al. [30] used this multibody model to estimate the loads and kinematic forces acting on the hip joint during walking and modelling with these inputs the deformation of an acetabular cup. Our study reported here focused on the influence that the radial clearance, namely the difference between the inner radius of the cup (R_c) and the radius of the femoral head (R_h)—see Figure 1—has on the pressure and deformation of the acetabular cup in a hard-on-soft bearing.

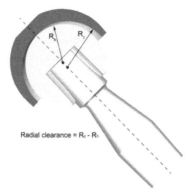

Figure 1. Schematization of the radial clearance between the inner surface of the acetabular cup (grey component in the picture) and the outer surface of the femoral head (pink component in the picture).

2. Materials and Methods

2.1. Finite Element Analysis (FEA) Model

Workbench Ansys® Software (v. 18.1, ANSYS Inc., Canonsburg, PA, USA) was used for the realization of the finite elements model. The model considered the hard-on-soft bearing configuration, typical of the most widespread hip implants [31]. To solve the contact problem, the solution algorithm must establish a relationship between the interacting surfaces, preventing the interpenetration. The software enforces contact compatibility through a penalty-based formulation that in this study is the augmented Lagrange, with a program-controlled penetration tolerance. With this formulation, a direct proportionality between the applied force and the penetration, where the coefficient is a contact stiffness, is established. Furthermore, the force is augmented by an extra term, making this method less sensitive to the magnitude of the contact stiffness. An asymmetric behavior was chosen for the contact, where the target body was the head and the contact body the cup, as all the resulting data belong to the contact side.

The simulations were performed comparing the effect of different clearances ranging from zero to 0.5 mm, namely 0, 0.05, 0.25 and 0.5 mm. The comparisons were in terms of magnitude and distribution of the contact pressure and of the deformation at the interface. As the sliding surfaces play a key role in the tribological performance of hip implants, the presence of the frictional force was considered. The coefficient of friction was obtained from experimental activities [32,33], considering a dry contact, and corresponded to 0.13. Even though the surface roughness could influence the friction and wear of a joint [34,35], the ease of computation and the very fine roughness—typical of actual hip replacements—led to a discharge of this aspect in the study. Contact area, pressure, and deformation were computed as a function of the angular displacements and dynamic loading calculated through a musculoskeletal model. Force components and orientations will be defined with respect to a pelvic reference frame that coincides with the true anatomic superior, anterior, and lateral directions.

Ultra-high-molecular-weight polyethylene (UHMWPE) type GUR 1050, was chosen for the acetabular cup whereas an infinitely rigid body was chosen as the femoral head, which is commonly realized in metal or ceramic. This choice is justified by the assumption that the site of wear interest is almost exclusively the soft body of the joint. Table 1 presents the main parameters of the acetabular cup material [36]. The presence of the pelvic bone has been neglected, since it has little influence on contact pressure [37]. The femoral head had a diameter of 28 mm, whereas the acetabular cup had a thickness of 5 mm. The cup was oriented with an anterior–posterior angle of 45°, and an inclination angle of 0°, these angles derived directly from the musculoskeletal model, from which forces and displacements were obtained. Therefore, it was not necessary to impose further inclination in the model.

Table 1. UHMWPE GUR 1050 properties.

Density	Young's Modulus	Poisson's Ratio	Bulk Modulus	Shear Modulus	Tensile Yield Strength	Tensile Ultimate Strength
(kg m^{-3})	(MPa)	(-)	(MPa)	(MPa)	(MPa)	(MPa)
930	690	0.46	1640	241	21	40

The finite element model is shown in Figure 2. A convergence study was performed to determine the optimum mesh aspect; contact pressure and cup deformation were used as performance metrics. For the cup the tetra patch conforming and tetra patch independent, the hex dominant quadrilateral/triangular and all quadrilateral, multizone hexagonal were compared; for the head, quadrilateral and triangular elements were compared. Regarding the head, quadratic elements were selected, for the cup a multizone hexagonal/prismatic hexagonal dominant meshing was chosen. The meshing of the entire model generated a total of 2012 elements and 3283 nodes, having 383 contact elements (Conta174 for the cup and Targe170 for the head). Elements of the mesh, having a minimum edge length of 53.7 mm, presented an average quality of 0.759 ± 0.152 and an average aspect ratio of 2.144.

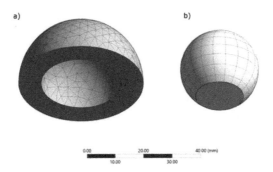

Figure 2. Meshing on (**a**) the acetabular cup and the (**b**) femoral head.

The input data obtained by the multibody system are in a local coordinates system, which follows the movements of the femoral head. As the software requires a global coordinates system, a conversion was required by using the rotational matrix. The three rotations around the axes are the Flexion/Extension (Z axes), Abduction/Adduction (X axes) and Inward/Outward (Y axes) (see Figure 3).

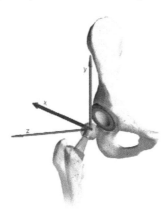

Figure 3. The three force directions.

2.2. Gait Cycles and Loads

In this study, we used musculoskeletal modeling software, to estimate loads acting on the hip joint during level walking. AnyBody Modelling System™ (AnyBody Technology A/S, Aalborg, Denmark) is a musculoskeletal modeling and simulation software. It combines the principles of the inverse dynamics with different algorithms to define the muscle recruitment and analyze the loads acting on different joints of the human body [38]. To calculate the joint forces, the kinematic data and ground reaction force must be known and used as inputs. These data are collected using special motion capture tools in the gait analysis laboratories [39], which measure a subject gait by means of cameras that monitor markers on the person's skin.

In inverse dynamics, the motion and the external loads on the body are known, and the aim is finding internal forces. Since not enough equilibrium equations are available to solve the problem, the evaluation of the muscle forces is made possible by the so-called redundancy problem. The working method of the AnyBody Modelling System is amply defined in [38], stating that there is the need to use an algorithm to find the activation of the muscles to reproduce the function of the central nervous system. The resolution of the multibody problem is already described in a previous work [30].

3. Results

The force components and rotations, obtained from the multibody analysis, are shown in Figure 4, plotted along the gait cycle percentage.

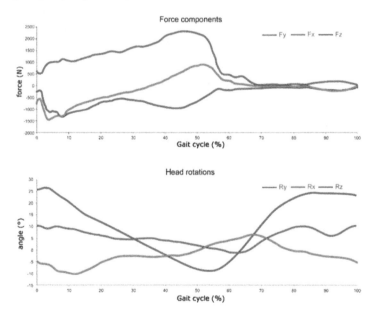

Figure 4. Forces and rotations obtained from multibody.

In Figure 5, pressure distribution, on the sliding surface of the acetabular cup during different steps are shown. Along with the cycle progressing, it is possible to observe the distributions of pressure and its magnitude. The rotation of the head (suppressed in the images for a better understanding) varies with the position of the highest value of pressure. When the pressure reaches its maximum values along the cycle, its major magnitude is in the edge zone of the cup.

In Figures 6 and 7 are the comparisons of the radial clearance on the pressure and the deformation, expressed in terms of the maximum value found on the inner surface of the cup in each timeframe.

From these images it is possible to highlight the divergences due to the clearance and the two peaks along the cycle. The presence of the two moments of high pressure and deformation already found in the previous analysis is clear.

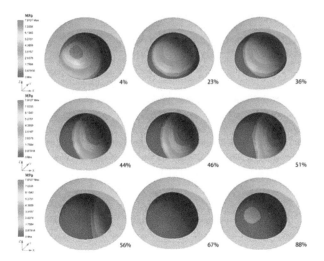

Figure 5. Pressure of the cup in several steps, as expresses in term of gait cycle percentage below each image.

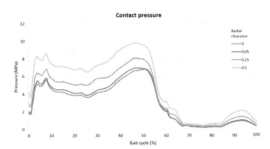

Figure 6. Maximum values of the pressure on the contact surface for different radial clearances.

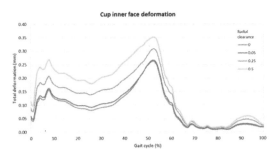

Figure 7. Maximum values of the deformation on the inner surface of the acetabular cup for different radial clearances.

Table 2 presents the highest values along the gait cycle found for pressure and deformation, comparing the radial clearances. These values were all located around the middle of the gait cycle, i.e., nearly 50%.

Table 2. Maximum values along the gait cycle of the pressure and deformation on the inner face of the cup.

Radial clearance	0	0.05	0.25	0.5
Maximum pressure (MPa)	6.82	7.00	8.12	9.84
Maximum deformation (mm)	0.26	0.27	0.31	0.35

4. Discussion

The increasing number of tribology studies to analyze polyethylene wear in total hip arthroplasty confirms the need for improved understanding and new solutions to avoid the failure of an implant due to polyethylene wear. Wear tests are performed on biomedical materials to solve or reduce failures or malfunctions due to material loss. It is well known that pre-clinical validation is considered as an extension in the risk analysis task, and wear tests in prosthetic hip implants are necessary in order to acquire quality control and further knowledge about the tribological processes in joint prostheses [7]. The "gold standard" accepted preclinical method to evaluate the wear performance of a hip implant design in the laboratory is performed using a joint simulator that simulates the physiological loadings and movements of the patients; these machines give important outcomes about the expected behavior of a hip implant in clinical use [7,30]. FEA methodology has been used for many years as a numerical technique to solve engineering problems related to stress and strain analysis of static or dynamic loaded contacts. The technique incorporates the principles of Newtonian mechanics to model and animate the realistic behavior of such contacts [40,41]. In addition, experimental studies are currently used, but they are quite expensive and time-consuming. In addition, these studies can analyze only limited configurations and load conditions. Some authors [42–45] used different protocols than those recommended by the ISO standard for hip and knee joint simulation [14,46,47] considering highly demanding activities such as stair climbing, chair sitting, squats, etc. Unfortunately, all these experimental approaches are time-consuming and very expensive. For these reasons, the use of computational modeling is expanding also in this orthopedic field, but unfortunately, few published papers present validated wear models to use.

The aim of this study was to investigate the influence of the radial clearance on the pressure and deformation of an acetabular cup made of UHMWPE. The investigation based its input data on a musculoskeletal multibody model, solved for a walking cycle. The results showed the behavior of the inner face of the cup, which is subjected to a dynamic pressure, changing magnitude and orientation along the gait cycle. In detail, the side zones of the cup are subjected to an elevated pressure during the first and last touching moments of the feet on the ground, when there is the impact and the release. These instants are known as heel strike and toe off, and their occurrence corresponds to an elevated bending in flexion.

The values of pressure find their agreement with previous studies, like the one performed by Liu et al. [16], where the maximum pressure corresponded to 8 MPa. The study on wear performed by Mattei et al. [48] showed a contact pressure of 8.2 MPa in correspondence with the maximum load. In the work of Matsoukas et al. [49] the highest values were almost 8 MPa and 12 MPa for the von Mises pressure and the contact pressure, respectively. These values were found in correspondence with the second peak of the stance phase, plus the pressure distribution concentrates in the edge zone of the cup. Maxian et al. [50] found a similar distribution of pressure reaching a maximum value of 9 MPa. The small differences that we found can be attributed to the dynamic loading retrieved from the musculoskeletal model, which differ from the standardized loads used in the literature. It is believed that before improving and making a more complex algorithm to find wear in silico, taking also into account realistic lubrication models, there is a need to apply more flexible and reliable loads as used in this study.

The comparison across the different radial clearance values highlighted that the there is a sensible divergence in term of pressure and deformation. From the 0 value of radial clearance to the highest, 0.5, there is a difference of 44% and 35% in terms of maximum pressure and deformation, respectively. This increment can be attributed to the modification of the contact distribution. In fact, when two convex–concave bodies pressing against each other are in perfect compliance the load is distributed across a wider area than if the same two bodies differ in dimension. Having a highly conformal contact means a wide area where forces are distributed and therefore a smaller pressure amplitude. As a consequence of a less intense pressure, the maximum values of the deformation are also smaller.

There is a very small variation between the two close cases of no clearance and 0.05 mm. In detail, the difference of these conditions is limited to the 3% of the maximum pressure and 4% of the maximum deformation. This variation is so little that it can be discarded in a first approximation, but even such a small difference can be relevant when lubrication comes into play. Lubrication in fact is dependent on different factors such as the viscosity of the lubricant, the finishing of the surfaces, and the shape of them. Research has demonstrated that lubricated hip joints could be modelled as a ball-in-socket or ball-on-plane equivalent configuration in boundary, mixed or hydrodynamic/elastohydrodynamic lubrication regimes. In this configuration, small radial clearances may benefit from lubrication phenomena by increasing minimum film thickness. Moreover, the greater entraining velocity in correspondence with low clearances could favor the hydrodynamic lubricating fluid film with a consequential positive effects on wear [24,51–54].

Obviously, our study has a number of limitations. The first limitation is due to the fact that our model was not validated experimentally on an in vitro simulation but it was validated only by comparison with international literature. It is stressed, however, that our model is in agreement with other in silico models. The second limitation is that our model was developed for a diameter of 28 mm, further studies are necessary to find out if there are differences with other dimensions. The model considered a single orientation of the cup, which is believed to be valid as stated by other studies [22], but limits its application to the specific case. Different cup inclination angles can influence the contact zone, as reported in the work of Hua et al. [55]: By increasing the cup inclination "the microseparation distances required to generate edge loading decreased". Other factors, such as additional effects of contact mechanics, or mechanics associated with three-body wear, temperature, and lubrication were not incorporated in this model and remain to be modelled.

5. Conclusions

The proposed finite element model allowed us to investigate the tribo-mechanical behavior of a hip implant subjected to dynamic loads derived from a multibody musculoskeletal model. The study focused on the differences found in pressure and deformation, in terms of distribution and amplitude, by varying the radial clearance in the range of 0–0.5 mm. The results highlighted the following main findings:

- along the walking cycle, the heel strike and the toe off are the instants of highest pressure, against these points the maximum pressure is located on the lateral side of the cup;
- radial clearance plays a substantial role on pressure and deformation magnitude across the gait cycle: the higher the clearance, the higher the pressure. Nevertheless, reducing the clearance to the smallest possible value is believed to affect the bearing lubrication;
- since radial clearance plays a key role in the lubricating phenomena, the proposed finite element model will be useful to exploit in the complete lubrication modelling of the joint toward a more and more realistic assessment of the in silico wear, accounting also for different radial clearances.

Author Contributions: S.A. and A.R. conceived and designed the investigation; S.A. developed the scientific framework; A.R. performed the multibody dynamical analysis; A.R. and M.M. developed the FEM model; S.A. and

Materials **2018**, *11*, 1282

M.M. selected the interest cases and analyzed the calculated data; S.A. and A.R. discussed the results; S.A., M.M. and A.R. wrote the paper.

Funding: This work was partially supported by the Italian Program of Donation for Research "5 per mille", year 2014, and by FARB 2017 obtained from University of Salerno.

Acknowledgments: The authors thank Simone Di Stasio for his key contribution to finite element modelling and Luigi Lena (Rizzoli Orthopaedic Institute) for his valuable help with the pictures.

Conflicts of Interest: The authors declare no conflicts of interest.

References

1. Affatato, S.; Freccero, N.; Taddei, P. The biomaterials challenge: A comparison of polyethylene wear using a hip joint simulator. *J. Mech. Behav. Biomed. Mater.* **2016**, *53*, 40–48. [CrossRef] [PubMed]
2. Affatato, S.; Bersaglia, G.; Emiliani, D.; Foltran, I.; Taddei, P.; Reggiani, M.; Ferrieri, P.; Toni, A. The performance of gamma- and EtO-sterilised UHMWPE acetabular cups tested under severe simulator conditions. Part 2: Wear particle characteristics with isolation protocols. *Biomaterials* **2003**, *24*, 4045–4055. [CrossRef]
3. Affatato, S.; Bersaglia, G.; Rocchi, M.; Taddei, P.; Fagnano, C.; Toni, A. Wear behaviour of cross-linked polyethylene assessed in vitro under severe conditions. *Biomaterials* **2005**, *26*, 3259–3267. [CrossRef] [PubMed]
4. Grillini, L.; Affatato, S. How to measure wear following total hip arthroplasty. *Hip Int.* **2013**, *23*, 233–242. [CrossRef] [PubMed]
5. Affatato, S.; Zavalloni, M.; Spinelli, M.; Costa, L.; Bracco, P.; Viceconti, M. Long-term in-vitro wear performance of an innovative thermo-compressed cross-linked polyethylene. *Tribol. Int.* **2010**, *43*, 22–28. [CrossRef]
6. Affatato, S.; Testoni, M.; Cacciari, G.L.; Toni, A. Mixed oxides prosthetic ceramic ball heads. Part 2: Effect of the ZrO_2 fraction on the wear of ceramic on ceramic joints. *Biomaterials* **1999**, *20*, 971–975. [CrossRef]
7. Affatato, S.; Spinelli, M.; Zavalloni, M.; Mazzega-Fabbro, C.; Viceconti, M. Tribology and total hip joint replacement: Current concepts in mechanical simulation. *Med. Eng. Phys.* **2008**, *30*, 1305–1317. [CrossRef] [PubMed]
8. Dumbleton, J.H. *Tribology of Natural and Artificial Joints*; Elsevier: New York, NY, USA, 1981.
9. Petersen, D.; Link, R.; Wang, A.; Polineni, V.; Essner, A.; Sokol, M.; Sun, D.; Stark, C.; Dumbleton, J. The Significance of Nonlinear Motion in the Wear Screening of Orthopaedic Implant Materials. *J. Test. Eval.* **1997**, *25*, 239. [CrossRef]
10. Affatato, S.; Bersaglia, G.; Foltran, I.; Emiliani, D.; Traina, F.; Toni, A. The influence of implant position on the wear of alumina-on-alumina studied in a hip simulator. *Wear* **2004**, *256*, 400–405. [CrossRef]
11. Zanini, F.; Carmignato, S.; Savio, E.; Affatato, S. Uncertainty determination for X-ray computed tomography wear assessment of polyethylene hip joint prostheses. *Precis. Eng.* **2018**, *52*, 477–483. [CrossRef]
12. Affatato, S.; Valigi, M.C.; Logozzo, S. Wear distribution detection of knee joint prostheses by means of 3D optical scanners. *Materials (Basel)* **2017**, *10*, 364. [CrossRef] [PubMed]
13. Affatato, S.; Zanini, F.; Carmignato, S. Quantification of wear and deformation in different configurations of polyethylene acetabular cups using micro X-ray computed tomography. *Materials (Basel)* **2017**, *10*, 259. [CrossRef] [PubMed]
14. *ISO DIS 14242–1–Implants for Surgery–Wear of Total Hip-Joint Prostheses–Part 1: Loading and Displacement Parameters for Wear–Testing Machines and Corresponding Environmental Conditions for Test*; International Organization for Standardization: Geneva, Switzerland, 2012.
15. Gao, L.; Dowson, D.; Hewson, R.W. Predictive wear modeling of the articulating metal-on-metal hip replacements. *J. Biomed. Mater. Res. Part B Appl. Biomater.* **2017**, *105B*, 497–506. [CrossRef] [PubMed]
16. Liu, F.; Fisher, J.; Jin, Z. Computational modelling of polyethylene wear and creep in total hip joint replacements: Effect of the bearing clearance and diameter. *Proc. Inst. Mech. Eng. Part J J. Eng. Tribol.* **2012**, *226*, 552–563. [CrossRef]
17. Suh, J.-K.; Bai, S. Finite Element Formulation of Biphasic Poroviscoelastic Model for Articular Cartilage. *J. Biomech. Eng.* **1998**, *120*, 195. [CrossRef] [PubMed]

18. Töyräs, J.; Saarakkala, S.; Laasanen, M.S.; Töyräs, J.; Korhonen, R.K.; Rieppo, J.; Saarakkala, S.; Nieminen, M.T.; Hirvonen, J.; Jurvelin, J.S. Biomechanical properties of knee articular cartilage. *Biorheology* **2003**, *40*, 133–140.

19. Li, G.; Gil, J.; Kanamori, A.; Woo, S.L.-Y. A Validated Three-Dimensional Computational Model of a Human Knee Joint. *J. Biomech. Eng.* **1999**, *121*, 657. [CrossRef] [PubMed]

20. Amirouche, F.; Solitro, G.; Broviak, S.; Goldstein, W.; Gonzalez, M.; Barmada, R. Primary cup stability in THA with augmentation of acetabular defect. A comparison of healthy and osteoporotic bone. *Orthop. Traumatol. Surg. Res.* **2015**, *101*, 667–673. [CrossRef] [PubMed]

21. Ghosh, R. Assessment of failure of cemented polyethylene acetabular component due to bone remodeling: A finite element study. *J. Orthop.* **2016**, *13*, 140–147. [CrossRef] [PubMed]

22. Uddin, M.S.; Zhang, L.C. Predicting the wear of hard-on-hard hip joint prostheses. *Wear* **2013**, *301*, 192–200. [CrossRef]

23. Huiskes, R.; Chao, E.Y.S.; Crippen, T.E. Parametric analyses of pin-bone stresses in external fracture fixation devices. *J. Orthop. Res.* **1985**, *3*, 341–349. [CrossRef] [PubMed]

24. Smith, S.; Dowson, D.; Goldsmith, A. The effect of diametral clearance, motion and loading cycles upon lubrication of metal-on-metal total hip replacements. *Proc. Inst. Mech. Eng. Part C* **2001**, *215*, 1–5. [CrossRef]

25. Felippa, C.A. Introduction to Finite Element Methods. 2004. Available online: https://www.colorado.edu/engineering/Aerospace/CAS/courses.d/IFEM.d/IFEM.Ch00.d/IFEM.Ch00.pdf (accessed on 19 July 2018).

26. Herrera, A.; Ibarz, E.; Cegoñino, J.; Lobo-escolar, A.; Puértolas, S.; López, E.; Gracia, L.; Herrera, A.; Lobo-escolar, A.; Mateo, J. Applications of finite element simulation in orthopedic and trauma surgery. *World J. Othop.* **2012**, *3*, 25–41. [CrossRef] [PubMed]

27. Maxian, T.A.; Brown, T.D.; Pedersen, D.R.; Callaghan, J.J. Adaptiive finite element modeling of long-term polyethylene wear in total hip arthroplasty. *J. Orthop. Res.* **1996**, *14*, 668–675. [CrossRef] [PubMed]

28. Hu, C.; Liau, J.; Lung, C.; Huang, C.; Cheng, C. A two-dimensional finite element model for frictional heating analysis of total hip prosthesis. *Mater. Sci. Eng. C* **2001**, *17*, 11–18. [CrossRef]

29. Sfantos, G.K.; Aliabadi, M.H.Ã. Total hip arthroplasty wear simulation using the boundary element method. *J. Biomech.* **2007**, *40*, 378–389. [CrossRef] [PubMed]

30. Ruggiero, A.; Merola, M.; Affatato, S. Finite element simulations of hard-on-soft hip joint prosthesis accounting for dynamic loads calculated from a Musculoskeletal model during walking. *Materials (Basel)* **2018**, *11*, 574. [CrossRef] [PubMed]

31. Affatato, S.; Ruggiero, A.; Merola, M. Advanced biomaterials in hip joint arthroplasty. A review on polymer and ceramics composites as alternative bearings. *Compos. Part B Eng.* **2015**, *83*, 276–283. [CrossRef]

32. Ruggiero, A.; D'Amato, R.; Gómez, E.; Merola, M. Experimental comparison on tribological pairs UHMWPE/TIAL6V4 alloy, UHMWPE/AISI316L austenitic stainless and UHMWPE/AL$_2$O$_3$ ceramic, under dry and lubricated conditions. *Tribol. Int.* **2016**, *96*, 349–360. [CrossRef]

33. Ruggiero, A.; D'Amato, R.; Gómez, E. Experimental analysis of tribological behavior of UHMWPE against AISI420C and against TiAl6V4 alloy under dry and lubricated conditions. *Tribol. Int.* **2015**, *92*, 154–161. [CrossRef]

34. Affatato, S.; Ruggiero, A.; De Mattia, J.S.; Taddei, P. Does metal transfer affect the tribological behaviour of femoral heads? Roughness and phase transformation analyses on retrieved zirconia and Biolox® Delta composites. *Compos. Part B Eng.* **2016**, *92*, 290–298. [CrossRef]

35. Affatato, S.; Ruggiero, A.; Merola, M.; Logozzo, S. Does metal transfer differ on retrieved Biolox® Delta composites femoral heads? Surface investigation on three Biolox® generations from a biotribological point of view. *Compos. Part B Eng.* **2017**, *113*, 164–173. [CrossRef]

36. Laurian, T.; Tudor, A. Some Aspects Regarding the Influence of the Clearance on the Pressure Distribution in Total Hip Joint Prostheses. In Proceeding of the National Tribology Conference RotTrib03, Annals, Galati, Romania, 24–26 September 2003.

37. Barreto, S.; Folgado, J.; Fernandes, P.R.; Monteiro, J. The Influence of the Pelvic Bone on the Computational Results of the Acetabular Component of a Total Hip Prosthesis. *J. Biomech. Eng.* **2010**, *132*, 54503. [CrossRef] [PubMed]

38. Damsgaard, M.; Rasmussen, J.; Christensen, S.T.; Surma, E.; de Zee, M. Analysis of musculoskeletal systems in the AnyBody Modeling System. *Simul. Model. Pract. Theory* **2006**, *14*, 1100–1111. [CrossRef]

39. Vaughan, C.; Davis, B.L.; O'Connor, J.C. Dynamics of Human Gait. Available online: https://isbweb.org/data/ (accessed on 6 April 2018).

40. Prati, E.; Freddi, A.; Ranieri, L.; Toni, A. Comparative fatigue damage analysis of hip-joint prostheses. *J Biomech.* **1982**, *15*, 808. [CrossRef]

41. Huiskes, R.; Chao, E.Y.S. A survey of finite element analysis in orthopaedic biomechanics: The first decade. *J. Biomech.* **1983**, *16*, 385–409. [CrossRef]

42. Abdel-Jaber, S.; Belvedere, C.; Leardini, Al.; Affatato, S.; Abdel Jaber, S.; Belvedere, C.; Leardini, Al.; Affatato, S. Wear simulation of total knee prostheses using load and kinematics waveforms from stair climbing. *J. Biomech.* **2015**, *48*, 3830–3836. [CrossRef] [PubMed]

43. Abdel-Jaber, S.; Belvedere, C.; De Mattia, J.S.; Leardini, A.; Affatato, S. A new protocol for wear testing of total knee prostheses from real joint kinematic data: Towards a scenario of realistic simulations of daily living activities. *J. Biomech.* **2016**, *49*, 2925–2931. [CrossRef] [PubMed]

44. Battaglia, S.; Belvedere, C.; Jaber, S.A.; Affatato, S.; D'Angeli, V.; Leardini, A. A new protocol from real joint motion data for wear simulation in total knee arthroplasty: Stair climbing. *Med. Eng. Phys.* **2014**, *36*, 1605–1610. [CrossRef] [PubMed]

45. Affatato, S. Towards wear testing of high demanding.pdf. *J Brazilian Soc. Mech. Sci. Eng.* **2018**, *40*, 260–266. [CrossRef]

46. *ISO-14243-2–Implants for Surgery–Wear of Total Knee-Joint Prostheses–Part 2: Methods of Measurement*; International Organization for Standardization: Geneva, Switzerland, 2000.

47. *ISO DIS 14243-3–Implants for Surgery–Wear of Total Knee Joint Prostheses–Part 1: Loading and Displacement Paramenters for Wear-Testing Machines with Displacement Control and Corresponding Environmental Conditions for Test*; International Organization for Standardization: Geneva, Switzerland, 2004.

48. Mattei, L.; Di Puccio, F.; Ciulli, E. A comparative study of wear laws for soft-on-hard hip implants using a mathematical wear model. *Tribol. Int.* **2013**, *63*, 66–77. [CrossRef]

49. Matsoukas, G.; Willing, R.; Kim, I.Y. Total Hip Wear Assessment: A Comparison Between Computational and In Vitro Wear Assessment Techniques Using ISO 14242 Loading and Kinematics. *J. Biomech. Eng.* **2009**, *131*, 41011. [CrossRef] [PubMed]

50. Maxian, T.A.; Brown, T.D.; Pedersen, D.R.; Callaghan, J.J. A sliding-distance-coupled finite element formulation for polyethylene wear in total hip arthroplasty. *J. Biomech.* **1996**, *29*, 687–692. [CrossRef]

51. Jin, Z.M.; Dowson, D. A full numerical analysis of hydrodynamic lubrication in artificial hip joint replacements constructed from hard materials. *Proc. Inst. Mech. Eng. Part C J. Mech. Eng. Sci.* **1999**, *213*, 355–370. [CrossRef]

52. Jalali-Vahid, D.; Jagatia, M.; Jin, Z.M.; Dowson, D. Prediction of lubricating film thickness in UHMWPE hip joint replacements. *J. Biomech.* **2001**, *34*, 261–266. [CrossRef]

53. Ruggiero, A.; Gòmez, E.; D'Amato, R. Approximate analytical model for the squeeze-film lubrication of the human ankle joint with synovial fluid filtrated by articular cartilage. *Tribol. Lett.* **2011**, *41*, 337–343. [CrossRef]

54. Ruggiero, A.; Gómez, E.; D'Amato, R. Approximate closed-form solution of the synovial fluid film force in the human ankle joint with non-Newtonian lubricant. *Tribol. Int.* **2013**, *57*, 156–161. [CrossRef]

55. Hua, X.; Li, J.; Wang, L.; Jin, Z.; Wilcox, R.; Fisher, J. Contact mechanics of modular metal-on-polyethylene total hip replacement under adverse edge loading conditions. *J. Biomech.* **2014**, *47*, 3303–3309. [CrossRef] [PubMed]

 *materials*

Article

Macrophage Biocompatibility of CoCr Wear Particles Produced under Polarization in Hyaluronic Acid Aqueous Solution

Blanca Teresa Perez-Maceda [1], María Encarnación López-Fernández [1], Iván Díaz [2], Aaron Kavanaugh [3], Fabrizio Billi [3], María Lorenza Escudero [2], María Cristina García-Alonso [2] and Rosa María Lozano [1,*]

[1] Cell-Biomaterial Recognition Lab., Department of Cellular and Molecular Biology, Centro de Investigaciones Biológicas (CIB-CSIC), Ramiro de Maeztu 9, 28040 Madrid, Spain; bpm@cib.csic.es (B.T.P.-M.); lfmarien@gmail.com (M.E.L.-F.)
[2] Department of Surface Engineering, Corrosion and Durability, Centro Nacional de Investigaciones Metalúrgicas (CENIM-CSIC), Avda. Gregorio del Amo 8, 28040 Madrid, Spain; ivan.diaz@cenim.csic.es (I.D.); escudero@cenim.csic.es (M.L.E.); crisga@cenim.csic.es (M.C.G.-A.)
[3] Department of Orthopaedic Surgery, David Geffen School of Medicine, University of California Los Angeles, Orthopaedic Hospital Research Center, 615 Charles E. Young Dr. South, Room 450A, Los Angeles, CA 90095, USA; akavanaugh@mednet.ucla.edu (A.K.); fabrizio.billi@gmail.com (F.B.)
* Correspondence: rlozano@cib.csic.es; Tel.: +34-918-373-112 (ext. 4208); Fax: 34-915-360-432

Received: 15 March 2018; Accepted: 2 May 2018; Published: 8 May 2018

Abstract: Macrophages are the main cells involved in inflammatory processes and in the primary response to debris derived from wear of implanted CoCr alloys. The biocompatibility of wear particles from a high carbon CoCr alloy produced under polarization in hyaluronic acid (HA) aqueous solution was evaluated in J774A.1 mouse macrophages cultures. Polarization was applied to mimic the electrical interactions observed in living tissues. Wear tests were performed in a pin-on-disk tribometer integrating an electrochemical cell in phosphate buffer solution (PBS) and in PBS supplemented with 3 g/L HA, an average concentration that is generally found in synovial fluid, used as lubricant solution. Wear particles produced in 3 g/L HA solution showed a higher biocompatibility in J774A.1 macrophages in comparison to those elicited by particles obtained in PBS. A considerable enhancement in macrophages biocompatibility in the presence of 3 g/L of HA was further observed by the application of polarization at potentials having current densities typical of injured tissues suggesting that polarization produces an effect on the surface of the metallic material that leads to the production of wear particles that seem to be macrophage-biocompatible and less cytotoxic. The results showed the convenience of considering the influence of the electric interactions in the chemical composition of debris detached from metallic surfaces under wear corrosion to get a better understanding of the biological effects caused by the wear products.

Keywords: polarization; CoCr alloy; wear particles; hyaluronic acid; macrophages biocompatibility

1. Introduction

Macrophages are cells involved in inflammatory processes [1]. All orthopedic biomaterials may induce a biologic host response to generated wear debris, which is strictly dependent on the nature of the debris. Metal wear particles and metal ions from prosthetic devices may induce a cascade of adverse cellular reactions that may include inflammatory complications, macrophage activation, bone resorption, and, although rarely, neoplasia [2,3]. In this context, macrophages play a decisive role in the hostile inflammatory reactions that can lead to implant loosening and failure.

Implanted metal surfaces in biological environments are exposed to cells and to physiological milieu interacting between them, an interaction that affects both the cells and the metallic surface. Implanted metallic materials, such as CoCr alloys, undergo dissolution and formation of a passive film that is affected by factors such as pH, ions present in the physiological medium, temperature, and biopotentials. Biopotentials are natural electrical properties that control the normal growth and development of different types of cells and tissues [4,5]. When a tissue is injured, its potentials undergo alterations to the normal potential of intact tissue [6,7]. Both biopotentials and injury potentials are found in bone and these potentials induced between injured and intact tissues persist until the tissue heals. Potentials in injured tissue can span over hundreds of microns and are generated by electric fields or ions flowing through the injured tissue [8,9] with a range of 10–100 mV/cm [10]. Assuming the resistivity of soft tissues to be 100 Ω cm [9,11], the resulting current density is in the 1–100 $\mu A/cm^2$ range [8,12]. Fukada and Yasuda had already described in 1957 the piezoelectric nature of the bone tissue [13]. Endogenous electrical properties of bone may play a role in the feedback mechanism of bone remodeling and development [14,15]. In vivo, these electrical signals work in collaboration to provide the correct environment for normal bone growth and development, but can be disrupted or altered by an injury after a trauma and during the healing process. Moreover, the resulting voltage gradients may induce modifications in the electrochemical potential of metallic implants and consequently may affect their surface properties.

Díaz et al. [16] recently characterized the CoCr alloy oxide films in a phosphate buffer solution containing 3 g/L of hyaluronic acid, the approximate concentration found in the synovial fluid of healthy joints [17], and under potentials with current density similar to those reported for injured tissues (1–100 $\mu A/cm^2$). Potentiostatic pulses applied during the growth of the CoCr oxide film produced a modification of the film that affected its chemical composition, thickness, and structure compared to the passive film formed in air [16]. These modifications induced surface heterogeneities at the atomic scale, geometric irregularities, such as nano-roughness, and a variation of the oxide composition [16]. Moreover, application of potentials of 0.7 V vs. Ag/AgCl induced changes in the oxide layer with the formation of 10–50 nm diameter nanopores, uniformly distributed along the surface and an increase in Cr (VI) and Mo (VI) concentration [16].

Despite the presence of the passive film, metals are susceptible to corrosion, particularly in aqueous environments, which may affect the surrounding tissue. Corrosion events generate electrical currents due to electron transfer from ions in the solution to the metallic surface where reactions are occurring. Wear-corrosion phenomena and micromotion or fretting-corrosion mechanically removes material, including the passive film, causing continuous activation/repassivation cycles [18]. These continuous and dynamic processes not only weaken the surface performance but also lead to an increase in the debris around the implant. Wear debris is considered one of the main factors responsible for aseptic loosening of orthopedic endoprostheses [19,20]. Implant failure due to aseptic loosening, or osteolysis, may result from the release of wear debris or electrochemical ions generated during corrosion events [20–22].

From the electrochemical point of view, on the metallic surfaces of implants, the breakdown of the passive film under the wear-corrosion process causes a drastic decrease in the open circuit potential of the metal towards negative potentials, i.e., from the passive to active state. This situation can suppose a polarization of about 500–700 mV with respect to the original open circuit potential. The change from the passive to active state can be induced mechanically under wear and electrochemically applying anodic polarization on the tribological system. Several researchers have studied the wear corrosion processes by application of anodic potentiodynamic polarization under wear processes [23,24].

The object of this paper was to evaluate the biocompatibility of particles produced during wear-corrosion assays of a CoCr alloy at potentiodynamic range to cover a wide polarization window on the samples. The hyaluronic acid, the lubricant component of the synovial liquid, was selected as the electrolyte for the generation of wear particles in conditions that represent more closely the prosthesis environment. Since macrophages are the main cells involved in the primary response to foreign bodies, cytotoxicity and biocompatibility of the wear particles were evaluated using these cells, measuring lactate dehydrogenase and mitochondrial activity, respectively.

2. Results and Discussion

2.1. Wear-Corrosion Tests

The interaction of physiological fluids with the bearing surfaces of hip implants is of great importance in the research of artificial joint lubrication, although this study has been so far little explored.

The effect of sliding of the alumina ball on the HCCoCr alloys is clearly shown in the drastic change in the open circuit potential. As an example, Figure 1 shows the change in the open circuit potential when the HCCoCr surfaces in PBS supplemented with 3 g/L HA (PBS-HA) are subjected to wear. The open circuit potential without wear was around −0.25 V versus Ag/AgCl, decreasing sharply when the alumina ball (pin) started the circular movement under 5 N load at 120 rpm. At this moment, the open circuit potential decreased until achieving values of about −0.55 V vs. Ag/AgCl, i.e., about 300 mV, and remained constant until the end of the test. The reduction in the potential value towards more negative values indicates that the HCCoCr surface becomes electrochemically active. This variation is due to the breakdown of the passive film under sliding, promoting the release of metallic ions and particles.

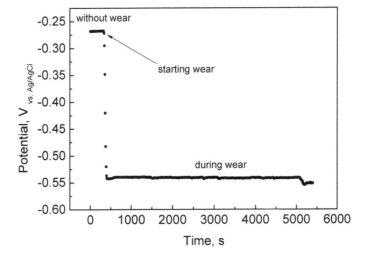

Figure 1. Open circuit potential of HCCoCr disks under wear. Measurement of the open circuit potential before and during the wear corrosion test of HCCoCr in PBS containing 3 g/L HA (PBS-HA).

Figure 2 panels a and b show the coefficient of friction (COF) for HCCoCr/alumina pair in PBS and PBS supplemented with 3 g/L HA (PBS-HA) during anodic potentiodynamic polarization and the anodic polarization curves drawn at 10 mV/min of HCCoCr in PBS and PBS containing 3 g/L HA under wear conditions (between point 1 and 2 in Figure 2a), respectively. The anodic polarization curve of HCCoCr in PBS-HA without wear has been also added in Figure 2b for comparative reasons. It can be seen that under sliding at the corrosion potential (before point 1 in Figure 2a), the COF was significantly higher in PBS than in PBS-HA. This result agrees with the hypothesis that the hyaluronic acid has a known lubricant role in the joint, acting as a shock absorber [25] and thus facilitating smooth joint movement by reducing friction between both surfaces. At the next stage (from point 1 to 2, in Figure 2a), the difference between both COF (in PBS and PBS-HA) remained, but higher fluctuations were detected. The fluctuations could be related to the continuous formation of hard particulate matter that enhances friction between both counterparts and decreases the friction when ejected from the track to surrounding areas where it accumulates (Figure 3). The load applied on the CoCr surfaces while sliding activates mechano-chemical reactions, causing not only the detachment of the passive film [26] but also bulk material resulting in an increase of COF. The hyaluronic acid in PBS maintains the lubricant effect during most of the wear corrosion tests.

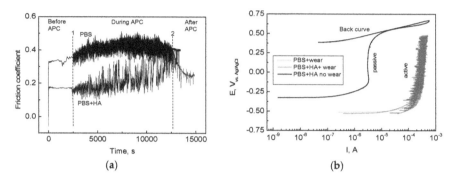

Figure 2. Friction coefficient of HCCoCr/alumina pair (**a**) and anodic polarization curves (APC) of HCCoCr disks (**b**) during wear corrosion tests in PBS and PBS containing 3 g/L HA (PBS-HA). Measurement of the friction coefficient before, during, and after application of anodic polarization current (APC). Anodic polarization curve for HCCoCr alloy in PBS-HA without wear is added for comparative analysis.

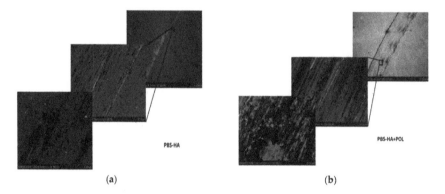

Figure 3. Secondary electron images of wear tracks on HCCoCr disks. Images by SEM of HCCoCr samples in PBS-HA under wear: (**a**) at the open circuit potential (PBS-HA) and (**b**) applying anodic potentiodynamic polarization (PBS-HA+POL).

As a consequence of mechanically assisted corrosion, the passive film on the HCCoCr surface was rapidly broken in both media, PBS and PBS-HA, producing an increase of approximately 3 orders of magnitude in current (Figure 2b) with respect to the anodic polarization curve without wear. Corrosion progresses on the wear track drawn by the sliding of alumina ball on the HCCoCr disks (Figure 3). Having in mind the wide passive region seen in the anodic polarization curve drawn without wear (Figure 2), the potential applied could be employed in forming rapidly the new oxide film. However, the sliding rate is quick enough to avoid the repassivation and formation of new protective chromium oxides. The constant value of the current density around 1 mA (three orders of magnitude higher than without wear) indicates that under these experimental conditions (5 N load and sliding rate of 120 rpm), the passive film is destroyed and remains in an active state until the end of the test.

Figure 3 shows the secondary electron (SE) images of the tracks of HCCoCr in PBS-HA after wear corrosion tests, at the corrosion potential (PBS-HA) and under anodic potentiodynamic polarization (PBS-HA+POL). In both cases (a) and (b), debris is accumulated in the immediate vicinity of the wear tracks, but the surface inside the track is especially altered when anodic potentiodynamic potential is applied. Figure 4 shows, as an example, the semiquantitative analysis taken by EDS of the three areas of interest in the HCCoCr alloy immersed in PBS-HA after wear corrosion under polarization (PBS-HA-POL): away from the track (spectrum 1), immediate vicinity (spectrum 2), and inside the track (spectrum 3). The most important feature found is the high % C content accumulated in the vicinity of the track. It means that the debris is mainly composed of C and O, the greatest proportion probably coming from the hyaluronic acid.

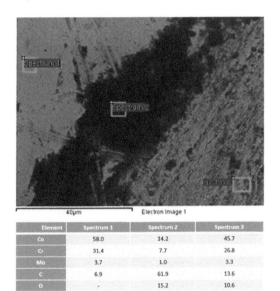

Element	Spectrum 1	Spectrum 2	Spectrum 3
Co	58.0	14.2	45.7
Cr	31.4	7.7	26.8
Mo	3.7	1.0	3.3
C	6.9	61.9	13.6
O	-	15.2	10.6

Figure 4. HCCoCr surface after wear corrosion applying anodic potentiodynamic polarization. Secondary electron image and EDS analyses away from the track, in the immediate vicinity, and inside the track in the HCCoCr surface in PBS-HA+POL.

The morphology and chemical characterization of the wear particles detached during wear corrosion tests revealed some interesting results. Figure 5 shows, as an example, the secondary electron image of wear particles collected from the tribocorrosion test in PBS containing 3 g/L HA and the semiquantitative analysis of some particles, identified from 1 to 6 and marked in blue color.

The statistical results of the effect of the corrosive medium and polarization applied in the wear corrosion tests on the chemical composition of the wear-detached particles collected appear in Table 1. In this table, three condition numbers assigned to 1, the PBS corrosive medium, 2, the PBS-HA corrosive medium without applying polarization, and 3, the PBS-HA corrosive medium applying polarization (PBS-HA-POL), have been considered. Mean, standard deviation, minimum and maximum value, C25 and C75, and median are shown.

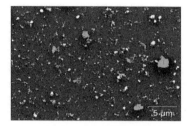

	C	O	Al	P	Cr	Co
	Wt.%	Wt.%	Wt.%	Wt.%	Wt.%	Wt.%
pt1	0	34.02	0	16.87	35.30	13.81
pt2	0	41.31	0	20.24	38.45	0
pt3	0	33.72	0	23.16	43.13	0
pt4	0	41.95	0	15.76	30.75	11.54
pt5	0	23.79	10.62	26.36	34.38	4.85
pt6	0	54.36	0	16.31	29.33	0

Figure 5. Secondary electron images of wear particles. Particles were collected from wear corrosion tests performed in PBS containing 3 g/L HA (PBS-HA) and deposited on silicon wafer to analyze the chemical composition of detached particles. Blue colors represent the particle where EDS has been performed and the particle number shown in pink is correlated to the pt shown in the table attached.

Table 1. Statistical analysis by Kruskal–Wallis test for the chemical composition (wt %) of wear particles detached during wear corrosion tests of CoCr samples in PBS (condition number 1), PBS+ 3 g/L HA (condition number 2) and under anodic potentiodynamic polarization (condition number 3), where n is the number of samples, mean is the average value, SD is the standard deviation, C25 is the value of the 25% of the data, C75 is the value of the 75% of the data, and p * is the significant difference at 0.95 confidence level.

	Condition Number	n	Mean	SD	Minimum	Maximum	C_{25}	Median	C_{75}	p *
	1	7	33.28	6.92	21.61	43.13	29.33	34.38	38.45	
Cr	2	24	28.53	8.73	14.06	48.82	21.28	27.52	33.66	0.034
	3	7	39.09	11.02	22.34	57.01	33.65	37.60	47.67	
	1	7	4.31	6.02	0.00	13.81	0.00	0.00	11.54	
Co	2	24	7.73	16.94	0.00	54.07	0.00	0.00	5.99	0.001
	3	7	36.08	17.59	15.14	58.64	17.57	32.12	51.48	
	1	7	0.00	0.00	0.00	0.00	0.00	0.00	0.00	
Mo	2	24	0.45	1.54	0.00	5.63	0.00	0.00	0.00	0.006
	3	7	2.71	2.60	0.00	5.48	0.00	4.00	5.45	
	1	7	18.59	5.02	11.46	26.36	15.76	16.87	23.16	
P	2	24	14.11	5.90	0.96	27.48	11.64	13.24	17.83	0.051
	3	7	9.40	7.30	1.30	19.57	1.89	8.79	16.10	
	1	7	1.52	4.01	0.00	10.62	0.00	0.00	0.00	
Al	2	24	0.18	0.68	0.00	3.17	0.00	0.00	0.00	0.123
	3	7	0.76	1.06	0.00	2.57	0.00	0.00	1.85	
	1	7	40.10	10.67	23.79	54.36	33.72	41.31	51.57	
O	2	24	41.82	17.72	3.52	72.84	34.64	49.54	51.69	0.002
	3	7	10.91	7.03	3.76	23.11	4.83	8.28	16.05	
	1	7	2.19	5.80	0.00	15.35	0.00	0.00	0.00	
C	2	24	6.98	7.49	0.00	19.94	0.00	6.39	14.16	0.022
	3	7	0.00	0.00	0.00	0.00	0.00	0.00	0.00	

* p-value in the Kruskal–Wallis test.

It can be seen that the particles are mainly composed of Co, Cr, Mo, P, C and O, with some traces of Al in some isolated particles. The Kruskal–Wallis test indicated that there are significant differences in the levels of Cr, Co, Mo, O and C, comparing the different conditions, i.e., depending on the composition of corrosive medium (PBS-condition 1 or PBS-HA-condition 2) and the application of polarization in wear corrosion tests (PBS-HA-POL, condition 3, and PBS-HA, condition 2). However, no significant differences in P and Al levels were obtained.

The results of the post hoc Mann–Whitney test used to determine which pairs differed among them are shown in Table 2. Cr levels are significantly higher in condition number 3 than 2 ($p = 0.021$). Co and Mo levels are significantly higher in condition number 3 than 1 and 2 ($p = 0.002$ and $p = 0.001$, $p = 0.025$ and $p = 0.002$, respectively). O levels are significantly lower in condition number 3 than 1 and 2 ($p = 0.002$ in both cases). C levels are significantly lower in condition number 3 than 2 ($p = 0.017$).

Table 2. Post hoc Mann–Whitney analysis to determine which pairs differed among them (condition numbers: 1-PBS, 2-PBS-HA, and 3-PBS-HA+POL).

	Comparison between Pairwise		
	p ** 1 vs. 2	p ** 1 vs. 3	p ** 2 vs. 3
Cr	0.119	0.277	0.021
Co	0.556	0.002	0.001
Mo	0.438	0.025	0.008
P	-	-	-
Al	-	-	-
O	0.508	0.002	0.002
C	0.098	0.317	0.017

** p-value in the Mann–Whitney test.

In summary, the statistical analysis confirmed that factors such as "composition of the corrosive medium" and "polarization applied" have an influence on the dependent variable chemical composition of the particles that is discussed immediately below.

The main significant effect of the addition of hyaluronic acid in the PBS to the wear particles detached is observed in the increase of the C content in the chemical composition of the particles. In both media (PBS, condition 1, and PBS-HA, condition 2), particles are mainly composed of Cr and O, followed by P and some Co. This chemical composition can be directly linked to the detachment of the native passive film during the wear corrosion test.

It has been proven by XPS (data not shown) that the immersion of the HCCoCr surfaces in PBS-HA causes a decrease in the Co species in the passive film and the enrichment in chromium oxide where phosphorus is included. It has been reported in literature that phosphate is adsorbed upon freshly exposed metal at the same time that ions are released into the solution until the passive layer is formed, whose composition varies significantly depending upon the environment [27]. Lewis et al. established that the corrosion, especially when associated with mechanical wear, is controlled by phosphate anions that absorb or react with the Co and Cr dissolution products. This promotes the formation of a mixed composition of phosphates, hydroxides, and oxides originating from the bulk metal.

This means that most of the particles collected after the wear corrosion tests in PBS and PBS-HA come from the native passive film (whose thickness is about 5–7 nm) and are mainly composed of chromium oxide and phosphate.

With respect to applying polarization during the wear corrosion tests in PBS-HA (condition 3), this factor has an important effect on the chemical composition of the wear particles detached. In this condition, particles are mainly composed of Cr and Co, followed by O, P, and Mo. The main significant effect of the polarization is the significant enrichment in Co, Cr, and Mo in the chemical composition of the detached particles. In this case, the wear particles produced under anodic polarization increased the Co/Cr ratio (with a value of 0.9 in comparison with a value of 0.3 found in PBS-HA without polarization). As wear particles obtained without polarization, these particles also contained P, although in a low proportion (Table 1). It has been reported in the literature [16] that the potential applied on the HCCoCr induces a change in the chemical composition of the passive film. Díaz et al. established that the increase in polarization (from 0.5 to 0.7 V) induced the preferential dissolution of cobalt whereas chromium was concentrated in the surface oxide film [16]. The passive film grown at a potential of 0.5 V vs. Ag/AgCl (into the passive region of the anodic polarization curve) consisted predominantly of Cr_2O_3 and $Cr(OH)_3$. However, the oxidation at a potential of 0.7 V vs. Ag/AgCl caused the appearance of Cr (VI) in the passive film but Co was not increased. In the case of wear corrosion under anodic potentiodynamic polarization, the continuous sliding of the alumina ball on the HCCoCr surface did not allow the regeneration of the oxide film. Instead, an active state stimulated by polarization was induced on the surface where bulk material was directly exposed and detached to the electrolyte. Considering this situation, the results reveal that the anodic polarization on CoCr surfaces under wear-corrosion processes accelerated and induced the release of larger metallic particles with higher Co content coming from the base material.

2.2. Macrophage Cell Response

Macrophages are a primary immune cell type and the main cellular type involved in inflammatory processes [1] and in host response [28], so their biologic host response to wear particles generated from the implanted materials is of great interest.

Macrophage response to wear particles derived from the tribocorrosion assays was evaluated by measuring the effect on cell toxicity and respiratory activity.

Cytotoxicity induced by HCCoCr wear particles was analyzed by measuring LDH activity released from cells (Figure 6), whose levels increase upon plasma membrane damage, a sign of cell death [29]. As is shown in Figure 6, exposure of macrophages cultures to wear particles induced a degree of cytotoxicity that was mainly dependent on the conditions used during wear-corrosion assays and particle concentration. As shown in Figure 6 panel A, particles concentration of 0.5 mg/mL obtained in PBS produced almost 58% cytotoxicity, a percentage that was significantly reduced to almost 12% when wear particles were generated from tribocorrosion tests in the presence of 3 g/L of hyaluronic acid (PBS-HA), an effect that could indicate a protective role of the hyaluronic acid on the metallic surface under wear stress conditions (Table 3). Concentrations of 1 mg/mL of wear particles from the PBS test produced an increase in the macrophage cytotoxicity to almost 75%, a value elevated in comparison with the cytotoxicity induced by the wear particles obtained in PBS containing 3 g/L of HA, where cytotoxicity reached 14% (data not shown). No additional increase in the cytotoxicity was observed at higher concentrations of wear particles (2 mg/mL) generated in PBS as macrophages cytotoxicity appeared comparable to the one elicited by exposure to lower concentrations of particles (0.5 and 1 mg/mL), where approximately a 64% cytotoxicity was detected (data not shown).

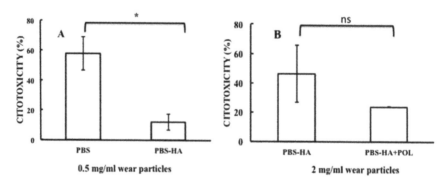

Figure 6. Macrophage cytotoxicity, measured as LDH activity, of cell cultures exposed for 72 h to HCCoCr wear particles. Panel A: Exposure of macrophages culture to 0.5 mg/mL wear particles. Particles were obtained in PBS and in PBS containing 3 g/L HA (PBS-HA). A *p* value of ≤ 0.05 was considered significant (*); Panel B: Exposure of macrophages culture to 2 mg/mL wear particles obtained in PBS containing 3 g/L HA with and without polarization application, PBS-HA+POL and PBS-HA, respectively. Experimental data were done as independent triplicate. Differences between data analyzed here were not significantly different (labeled as ns).

Table 3. Statistical analyses of cytotoxicity data. Mean differences of cytotoxicity effects between P2 (0.5 mg/mL) vs. P3 (0.5 mg/mL) and P3 (2 mg/mL) vs. P6 (2 mg/mL) were studied with Student's *t* tests (α = 0.05), respectively. (**a**) Wear particles obtained in PBS (P2) and in PBS containing 3 g/L HA (P3). (**b**) Wear particles obtained in PBS containing 3 g/L HA (P3) and in PBS containing 3 g/L HA with polarization application (P6).

(a)		
P2 (0.5 mg/mL) vs. P3 (0.5 mg/mL)		
Mean of P2	Mean of P3	*P* Value
57.84	12.24	0.015
(b)		
P3 (2 mg/mL) vs. P6 (2 mg/mL)		
Mean of P3 (2 mg/mL)	Mean of P6 (2 mg/mL)	*P* Value
46.18	23.9	0.248

Particles produced in PBS containing 3 g/L of HA at concentrations of 2 mg/mL elicited an increase in the macrophages cytotoxicity that reached almost 46% (Figure 6, panel B, PBS-HA). Although such an increase was higher than the one produced by particles concentrations of 0.5 and 1 mg/mL, which were 12% and 14%, respectively, it was reduced to 24% when polarization conditions characteristic of damaged tissue were applied (Figure 6, panel B, PBS-HA+POL). Although the statistical analysis of the data from Figure 6 panel B (Table 3) gave no significant differences between results analyzed here, the application of anodic polarization to HA aqueous solution seems to have important observable differences on the mean value of the cytotoxicity. This feature could be relevant and, for this reason, verification by other biocompatibility assays is required. With this purpose, the wear particles collected from the tribocorrosion assays of CoCr alloy in PBS-HA without and applying anodic polarization were tested on macrophages cultures by measuring the mitochondrial activity. It is well known that the mitochondrial activity measurement is directly proportional to the number of metabolically active cells in culture [29] constituting a measure of cell viability and biocompatibility. As it is shown with Figure 7 by white bars, wear particles collected in the PBS-HA produced a gradual and significant reduction in the mitochondrial respiratory response of macrophages. This result seemed to be directly related to the concentration of particles to which macrophages were exposed (Table 4). Nevertheless, no reduction in the mitochondrial respiratory activity was observed in macrophages exposed to wear particles generated when polarization was applied during wear-corrosion tests. No significant effects in respiratory activity were observed in the range of particles concentrations tested (Figure 7, black dotted bars, and Table 5). The results suggest that the polarization conditions in the wear-corrosion assays in PBS containing HA at the approximate concentration found in synovial fluid seem to be beneficial to macrophage viability and biocompatibility.

Table 4. Statistical analyses of mitochondrial respiratory activity. The effects of the particle (P), the concentration, and their interaction on the changes in respiratory activity were analyzed with a two-way analysis of variance. A *p* value of ≤ 0.05 was considered significant. Mean pairwise comparisons were computed with a Tukey's test (α = 0.05). All analyses were performed with the R software version 3.4.2 (R Core Team, Vienna, Austria, 2017).

ANOVA				
	Sum Sq	**Df**	**F Value**	**Pr (>F)**
P	5745.2	1	266.75	4.64×10^{-9}
conc	2191.39	2	50.87	2.76×10^{-6}
P × conc	2648.33	2	61.48	1.07×10^{-6}
Residuals	236.91	11	-	-

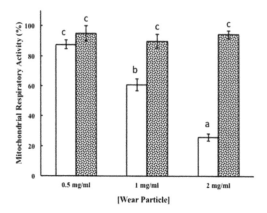

Figure 7. Mitochondrial respiratory activity of macrophages cell cultures exposed for 72 h to different doses of HCCoCr wear particles. Wear particles were obtained in PBS containing 3 g/L HA without (white bars) and with polarization (dotted black bars). Cell cultures were exposed to the following wear particles concentrations: 0.5, 1 and 2 mg/mL. Experiments were done as independent triplicate. Bars labeled with different letters show statistically significant differences and bars labeled with the same letter (c) show nonsignificant differences.

Table 5. As the interaction was significant, simple effects were compared. Means with the same letter are not significantly different.

P	Conc	Mean	Lower.CL	Upper.CL	Group
P3	2	25.7	17.1	34.3	a
P3	1	60.9	52.3	69.4	b
P3	0.5	87.5	77	98	c
P6	1	89.8	81.3	98.4	c
P6	2	94.3	85.8	102.9	c
P6	0.5	95	86.4	103.5	c

The dose-dependence effect on mitochondrial respiratory activity by particles detached in PBS-HA could be explained by the chemical composition of wear particles collected from wear-corrosion tests in this solution as a decrease in Co was observed, as well as an enrichment in chromium oxide, a compound with high toxicity [30] in comparison with the composition of PBS-HA-POL wear particles. The results suggest that polarization conditions applied to an HA aqueous solution at the approximate concentration found in synovial fluid produce changes in material tribocorrosion behavior inducing wear particles that seem to be beneficial to macrophage viability and biocompatibility. Data that could explain the higher biocompatibility of wear particles generated in PBS-HA+POL could be related to the fact that under these conditions, wear processes did not allow the regeneration of the oxide film. This event could determine the creation of an active state where CoCr base material was directly exposed to the electrolyte without enough time to build up the new oxide film that induced the release of metallic particles with higher Co content probably coming from the base material.

3. Materials and Methods

3.1. Material

A high carbon CoCr alloy (hereafter HCCoCr) that complies with ASTM F75 standard was used as material. HCCoCr composition is shown in Table 6. "Double heat-treated" disks, i.e., solution treatment (ST) followed by hot isostatically pressing (HIP), of 38 mm in diameter and 4 mm thickness, were obtained from BIOMET Spain Orthopaedic (Valencia, Spain). The sample preparation consisted of grinding on SiC paper, followed by mechanical polishing with 3 μm diamond paste.

Table 6. Chemical composition (wt %) of High Carbon CoCr alloy (HCCoCr).

	C	Co	Cr	Mo	Ni	S	P	Al	W	Mn	Fe	Si	N	Ti	Cu
HC	0.22	62	29.4	6.4	0.1	0.004	0.001	0.01	0.03	0.7	0.16	0.7	0.16	-	-

3.2. Wear-Corrosion Tests under Electrochemical Control

Wear-corrosion experiments were carried out on a pin-on-disk tribometer, and 6-mm diameter alumina ball pins were used as HCCoCr disk counterpart. The HCCoCr disks were 38 mm in diameter and 4 mm thick. Both disks and pins were previously washed with double distilled water and cleaned in an ultrasonic ethanol bath for 10 min. The alumina pins were placed in a pin plastic holder and fixed on the load cell. A low normal load of 5 N was applied on the counterpart. The working electrode motion was provided by a rotating motor at a rotation rate of 120 rpm that produced, at the end of the alumina ball, a circular wear track (5 mm in diameter) on the HCCoCr disk surface.

The tribometer configuration consisted of an integrated electrochemical cell (3-electrode cell) including the HCCoCr disk as working electrode, a ring-shaped Pt wire counter electrode and a saturated Ag/AgCl reference electrode. All the potentials of the HCCoCr disks during the wear corrosion tests were measured versus the reference electrode. Wear-corrosion tests were performed in Phosphate Buffer Solution (PBS) containing the following composition: 0.2 g/L KCl, 0.2 g/L KH_2PO_4, 8 g/L NaCl, and 1.150 g/L Na_2HPO_4 (anhydrous) and this PBS solution was supplemented with 3 g/L hyaluronic acid, the approximate concentration reported for the synovial fluid of healthy joints [17].

The wear-corrosion behavior was studied simultaneously measuring the friction coefficient and electrochemical parameters. The wear-corrosion tests were performed as follows (Figure 8): (a) before wear (no sliding) by the measurement of the corrosion potential for 10 min and (b) under sliding (at 120 rpm and 5 N load) in two different ways: one where a simultaneous measurement of the corrosion potential and the coefficient of friction (COF) were performed for 40 min without applying polarization and the other, applying anodic potentiodynamic polarization and simultaneous measurement of current and coefficient of friction (COF) for 200 min. The anodic potentiodynamic polarization was applied from the corrosion potential to a polarization of 1 V at a scanning rate of 10 mV/min, and back curve was drawn until reaching the corrosion potential. The back curve was also measured to analyze the repassivation ability of the HCCoCr alloy. For comparative purposes, the anodic potentiodynamic polarization without wear in PBS and PBS-HA was also measured. All experiments were carried out in triplicate.

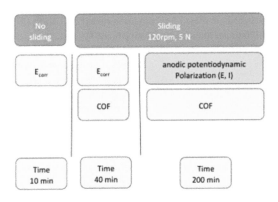

Figure 8. Schema of the experimental procedure of the wear-corrosion tests.

Surface characterization of the worn surface after wear-corrosion tests with polarization and without polarization was performed by profilometry and using a JEOL-6500F microscope equipped with a field Emission Gun (FEG) coupled to an Energy Dispersive X-ray (EDS) spectrometer. Secondary electron (SE) images were taken at 7.5 keV and EDS analysis was performed at 20 keV.

3.3. Isolation and Characterization of Particles

Debris from tribocorrosion tests performed in PBS, PBS containing 3 g/L hyaluronic acid (both without applying polarization), and in PBS supplemented with hyaluronic acid under anodic polarization was collected for the subsequent characterization. Metallic particles were isolated, purified, and characterized following the protocol developed by Billi et al. for metal particles [31]. This procedure allows exhaustive removal of organic and inorganic impurities from the metallic particles. To completely digest the hyaluronic acid, metal particles in PBS supplemented with 3 g/L HA were digested adapting the protocol developed by Kavanaugh et al. [32].

Wear-corrosion media (PBS and the digested hyaluronic acid solutions) containing metallic particles were rotated at 28 rpm in an orbital agitator for 24 h at room temperature to disperse the metal particles evenly before their isolation. The particles were then purified via density gradient centrifugation varying from $4446 \times g$ to $284,000 \times g$ (Beckman Optima L80 XP; Beckman Instruments, Fullerton, CA, USA) through multiple layers of denaturants and metal-selective high-density layers as was described [31,32]. This led to well-dispersed particles deposited onto a 5 mm × 5 mm featureless display silicon wafer (Ted Pella, Inc., Redding, CA, USA) coated with a monolayer of marine mussel glue (Cell-TakTM; BD Biosciences, San Jose, CA, USA). The silicon wafer was then coated with 10 Å iridium [31,32].

The morphology of metallic particles was studied with a field emission scanning electron microscope (FE-SEM) (Supra VP-40; Zeiss, Peabody, MA, USA) at a voltage of 15 kV and chemically analyzed by means of Energy-dispersive spectroscopy (EDS) analysis (Thermo Ultradry feature sizing system; Thermo Electron Scientific Instruments, Madison, WI, USA).

3.4. Macrophages Cell Cultures Assays

The biocompatibility of wear particles was tested in a mouse macrophage cell line (J774A.1) from DSMZ Human and Animal Cell Bank. Macrophages cell cultures were exposed to different concentrations of wear particles.

Wear particles obtained from the wear-corrosion tests were centrifuged and the particle pellet was weighted, UV sterilized for 15 min, and resuspended in sterile bidistillated water and maintained in aliquots at -20 °C until use. Wear particles, just before cell cultures assays, were thawed, resuspended by vigorously mixing with a vortex, and diluted at a concentration of 20 mg/mL in Dulbecco's Modified Eagle Medium (DMEM 41966; Gibco, BRL, Invitrogen, Thermofisher scientific, Paisley, UK) supplemented with 10% heat-inactivated fetal bovine serum (FBS; Gibco, BRL) and with a mixture of antibiotics (penicillin at 100 units/mL and streptomycin at 100 g/mL, Gibco, BRL), named as complete cell culture medium. A concentration of 20 mg/mL was used as stock solutions for the particles concentration tested in different cell assays. To assure a polydisperse distribution of the particles vigorous vortexing was applied in all experimental steps that required particles manipulation.

To evaluate the effect of HCCoCr particles on cell cultures, macrophages were seeded on 96-well culture plates at 75,000 cells/mL cell density in complete cell culture medium. A final volume of 100 μL of cell suspension in complete cell culture medium was added to each well of the 96-well plates. After 24 h in culture, cell media were removed and replaced by 100 μL of fresh complete cell culture medium containing the following concentrations of HCCoCr particles: 0, 0.5, 1 and 2 mg/mL. Cell cultures were maintained for 72 h in a cell culture chamber at 37 °C and 5% CO_2. Incubation time was selected based on the set-up of cell cultures assays for metallic particles studies carried out in the lab and is the most commonly used time point for cell viability studies [33]. Mitochondrial activity

(WST-1 assay) and plasma membrane damage (LDH assay) were used to evaluate the biocompatibility and cytotoxicity, respectively, as described below [29].

3.5. Mitochondrial Activity Measurement

Reduction of the WST-1 reagent (4-[3-4-iodophenyl)-2-(4-nitro-phenyl)-2H-5-tetrazolio]-1,3-benzene disulfonate (Roche Diagnostics GmbH, Mannheim, Germany)) was used to evaluate the effect of different concentrations of the HCCoCr wear particles on mitochondrial activity of macrophages cultures. The mitochondrial activity measurement is directly proportional to the number of metabolically active cells in culture. After 72 h in culture, 10 μL of the cell proliferation kit reagent WST-1 was added to each well containing 100 μL of fresh complete cell culture medium, and the mixture was incubated inside the cell culture incubator for 30 min. After incubation, 100 μL of each reaction mixture were transferred to a 96-well cell plate, and the absorbance of the samples was measured as differential absorbance, 415 nm minus 655 nm, in an iMark microplate absorbance reader (Bio-Rad, Hercules, CA, USA), using the absorbance given by complete cell culture medium as a blank. All experiments were carried out as independent triplicate.

3.6. Measurement of Lactate Dehydrogenase Activity

To measure and quantify the effect of HCCoCr wear particles on cell death and cell lysis, lactate dehydrogenase (LDH) activity was measured in the supernatants of cell cultures by an enzymatic assay using the Cytotoxicity Detection Kit[plus] (Roche Diagnostics GmbH, Mannheim, Germany). Supernatants were collected from cell culture after being exposed for 72 h to different HCCoCr particles concentrations and were centrifuged for 5 min at $1024 \times g$. The enzymatic assays were performed according to the LDH kit protocol provided by Roche Diagnostics (Mannheim, Germany). Complete cell culture medium was used as a control for absorbance baseline. LDH activity was measured based on differential absorbance, 490 nm minus 655 nm, in an iMark microplate absorbance reader (Bio-Rad, Hercules, CA, USA). LDH catalyzes the conversion of lactate to pyruvate, reducing NAD^+ to $NADH/H^+$, which is used by the catalyst to reduce a tetrazolium salt to a formazan salt, which is responsible for the change in absorbance at 490 nm. Quantification of LDH activity is used as an indicator of plasma membrane damage, as is a stable cytoplasmic enzyme present in all cells and rapidly release into the cell culture supernatant when the plasma membrane is damaged being a sign of cell death. The percentage of cytotoxicity is calculated taking as control a total cell lysate in the absence of any particles. The percentage cytotoxicity is calculated as described in the LDH kit protocol provided by Roche Diagnostics: Cytotoxicity (%) = [(exp. value − low control)/(high control − low control)] × 100; where experimental value (exp. value) corresponds to the absorbance of the treated sample in the study exposed to wear HCCoCr particles, low control is the absorbance from the untreated cell cultures with no particles that corresponds to spontaneous LDH released, and high control is the absorbance value obtained after total cell cultures lysis that corresponds to the maximum releasable LDH activity. The background absorbance corresponding to complete cell culture media was subtracted from the absorbance of all samples before cytotoxicity calculations. All experiments were carried out as independent triplicate.

3.7. Statistical Analysis of Data

3.7.1. Wear Particles Analysis Data

The experimental design used to determine the effect of two factors as the corrosive medium and the application of polarization on the dependent variable, that is, the chemical composition of the particles, was a 2^2 factorial design. In order to explain significant interaction, simple effects of one factor on the dependent variable at each single level of the other factor were computed. After this, simple effect pairwise comparisons were performed to detect levels of the second factor in which simple effects of the first factor on the dependent variable were significantly different.

Kruskal–Wallis [34] and Mann–Whitney [35] nonparametric tests were used to confirm the ANOVA results. A p-value < 0.05 was considered as significant. All the statistical analyses were performed with the Minitab® 17.1.0 software (Minitab Inc., State College, PA, USA) [36].

3.7.2. Biocompatibility Analysis Data

Mean differences on cytotoxicity effects between wear particles obtained in PBS (0.5 mg/mL) versus particles in PBS containing 3 g/L of hyaluronic acid (0.5 mg/mL) and between wear particles obtained in PBS containing 3 g/L HA (PBS-HA; 2 mg/mL) without versus with polarization application (PBS-HA+POL; 2 mg/mL) were studied with Student's t tests (α = 0.05), respectively.

The effects of the particles, the concentration, and their interaction on the changes in mitochondrial respiratory activity of macrophages were analyzed with a two-way analysis of variance. A p value of $\leq$0.05 was considered significant. Mean pairwise comparisons were computed with a Tukey's test (α = 0.05). Means with the same letter are not significantly different and means with different letters are significantly different.

All analyses were performed with the R software version 3.4.2 (R Core Team, Vienna, Austria, 2017) [37].

4. Conclusions

1. The wear particles collected after wear corrosion in PBS and PBS-HA were mainly composed of chromium oxide coming from the detachment of the passive film and phosphate adsorbed on the particle surface and/or adsorbed on the broken passive film.

2. Composition of the corrosive medium and polarization, applied to mimic the electrical interactions observed in living tissues, has an influence on the chemical composition of the particles. The wear particles detached after wear corrosion with polarization in PBS-HA have a chemical composition with a higher significant content of Cr and Co than those particles collected without polarization.

3. Biocompatibility in vitro assays here reported, measured by LDH release and mitochondrial respiratory activity, seem to indicate that particles from wear corrosion in PBS supplemented with 3 g/L of hyaluronic acid, an approximate concentration that is found in the synovial fluid of healthy joints, under anodic polarization produce in macrophages lower damage to the plasma membrane and are more biocompatible, most likely associated with particles chemical composition.

4. As more variables of the prosthesis environment are considered in in vitro assays to study cell-biomaterial interactions, as are the electric interactions, in order to have a closer view of the different processes that are taking place in vivo at the cell-biomaterial interface, a better knowledge of the biological consequences will be obtained.

5. Understanding these consequences of the electrical signals on the growth and development of cells and tissues should be applicable for the design of appropriate solutions and adequate treatments for orthopedic-bearing patients.

Author Contributions: Conceptualization, B.T.P.-M., M.L.E., M.C.G.-A., R.M.L.; Methodology, B.T.P.-M., M.E.L.F., I.D., A.K., F.B., M.L.E., M.C.G.-A., R.M.L.; Software, B.T.P.-M., M.L.E., M.C.G.-A., R.M.L., G.P., J.G.; Validation, B.T.P.-M., I.D., A.K., F.B., M.L.E., M.C.G.-A., R.M.L; Formal Analysis, B.T.P.-M., M.E.L.F., I.D., M.L.E., M.C.G.-A., R.M.L.; Investigation, B.T.P.-M., M.E.L.F., I.D., A.K., F.B., M.L.E., M.C.G.-A., R.M.L.; Resources, B.T.P.-M., M.L.E., M.C.G.-A., R.M.L.; Data Curation, B.T.P.-M., M.L.E., M.C.G.-A., R.M.L.; Writing-Original Draft Preparation, B.T.P.-M., M.L.E., M.C.G.-A., R.M.L.; Writing-Review & Editing, B.T.P.-M., M.L.E., M.C.G.-A., R.M.L.; Visualization, B.T.P.-M., M.L.E., M.C.G.-A., R.M.L.; Supervision, B.T.P.-M., M.L.E., M.C.G.-A., R.M.L.; Project Administration, B.T.P.-M., M.L.E., M.C.G.-A., R.M.L.; Funding Acquisition, B.T.P.-M., M.L.E., M.C.G.-A., R.M.L.

Funding: Financial support received through the MAT2015-67750-C3-2-R, MAT2015-67750-C3-1-R, MAT2011-29152-C02-01 and the MAT2011-29152-C02-02 projects from the Ministerio de Economía y Competitividad (MINECO/FEDER) from Spain.

Acknowledgments: Authors wish to thank Guillermo Padilla PhD G.P. (Bioinformatics and Biostatics facility at Centro de Investigaciones Biológicas, CIB-CSIC) and J. Garrido J.G. (U. Autonoma Madrid, Dept. Psicol. Social & Metodol., Fac. Psicol.) for technical assistance in the statistical analysis of macrophages data and wear particles.

Conflicts of Interest: There are no conflicts of interest to declare. The authors will receive no benefit of any kind either directly or indirectly.

References

1. Fujiwara, N.; Kobayashi, K. *Macrophages in Inflammation in Current Drug Targets—Inflammation & Allergy*; Zaenker, K.S., Ed.; Bentham Science Publisher: Berlin, Germany, 2005; pp. 281–286. [CrossRef]
2. Poggio, C.E. Plasmacytoma of the mandible associated with a dental implant failure: A clinical report. *Clin. Oral Implant. Res.* **2007**, *18*, 540–543. [CrossRef] [PubMed]
3. McGuff, H.S.; Heim-Hall, J.; Holsinger, F.C.; Jones, A.A.; O'Dell, D.S.; Hafemeister, A.C. Maxillary osteosarcoma associated with a dental implant: Report of a case and review of the literature regarding implant-related sarcomas. *J. Am. Dent. Assoc.* **2008**, *139*, 1052–1059. [CrossRef] [PubMed]
4. Ferrier, J.; Ross, S.M.; Kanehisa, J.; Aubin, J.E. Osteoclasts and osteoblasts migrate in opposite directions in response to a constant electrical-field. *J. Cell. Physiol.* **1986**, *129*, 283–288. [CrossRef] [PubMed]
5. Levin, M.; Thorlin, T.; Robinson, K.R.; Nogi, T.; Mercola, M. Asymmetries in H+/K+-ATPase and cell membrane potentials comprise a very early step in left-right patterning. *Cell* **2002**, *111*, 77–89. [CrossRef]
6. Becker, R.O.; Spadaro, J.A.; Marino, A.A. Clinical experiences with low intensity direct-current stimulation of bone-growth. *Clin. Orthop. Relat. Res.* **1977**, *124*, 75–83. [CrossRef]
7. Levin, M. Large-scale biophysics: Ion flows and regeneration. *Trends Cell Biol.* **2007**, *17*, 261–270. [CrossRef] [PubMed]
8. Lokietek, W.; Pawluk, R.J.; Bassett, C.A. Muscle injury potentials source of voltage in undeformed rabbit tibia. *J. Bone Jt. Surg. Br.* **1974**, *56*, 361–369. [CrossRef]
9. McCaig, C.D.; Rajnicek, A.M.; Song, B.; Zhao, M. Controlling cell behavior electrically: Current views and future potential. *Physiol. Rev.* **2005**, *85*, 943–978. [CrossRef] [PubMed]
10. Gittens, R.A.; Olivares-Navarrete, R.; Tannenbaum, R.; Boyan, B.D.; Schwartz, Z. Electrical implications of corrosion for osseointegration of titanium implants. *J. Dent. Res.* **2011**, *90*, 1389–1397. [CrossRef] [PubMed]
11. Faes, T.J.; van der Meij, H.A.; de Munck, J.C.; Heethaar, R.M. The electric resistivity of human tissues (100 Hz–10 MHz): A meta-analysis of review studies. *Physiol. Meas.* **1999**, *20*, R1–R10. [CrossRef] [PubMed]
12. Borgens, R.B.; Jaffe, L.F.; Cohen, M.J. Large and persistent electrical currents enter the transected lamprey spinal-cord. *Proc. Natl. Acad. Sci. USA* **1980**, *77*, 1209–1213. [CrossRef] [PubMed]
13. Fukada, E.; Yasuda, I. On the piezoelectric effect of bone. *J. Phys. Soc. Jpn.* **1957**, *12*, 1158–1162. [CrossRef]
14. Guzelsu, N.; Demiray, H. Electro-mechanical properties and related models of bone tissues—Review. *Int. J. Eng. Sci.* **1979**, *17*, 813–851. [CrossRef]
15. Rubinacci, A.; Black, J.; Brighton, C.T.; Friedenberg, Z.B. Changes in bioelectric potentials on bone associated with direct-current stimulation of osteogenesis. *J. Orthop. Res.* **1988**, *6*, 335–345. [CrossRef] [PubMed]
16. Díaz, I.; Martínez-Lerma, J.F.; Montoya, R.; Llorente, I.; Escudero, M.L.; García-Alonso, M.C. Study of overall and local electrochemical responses of oxide films grown on CoCr alloy under biological environments. *Bioelectrochemistry* **2017**, *115*, 1–10. [CrossRef] [PubMed]
17. Hui, A.Y.; McCarty, W.J.; Masuda, K.; Firestein, G.S.; Sah, R.L. A systems biology approach to synovial joint lubrication in health, injury, and disease. *Wiley Interdiscip. Rev. Syst. Biol. Med.* **2012**, *4*, 15–37. [CrossRef] [PubMed]
18. Hodgson, A.W.E.; Kurz, S.; Virtanen, S.; Fervel, V.; Olsson, C.-O.A.; Mischler, S. Passive and transpassive behavior of CoCrMo in simulated biological solutions. *Electrochim. Acta* **2004**, *49*, 2167–2178. [CrossRef]
19. Bitar, D.; Parvizi, J. Biological response to prosthetic debris. *World J. Orthop.* **2015**, *6*, 172–189. [CrossRef] [PubMed]
20. Jacobs, J.J.; Roebuck, K.A.; Archibeck, M.; Hallab, N.J.; Glant, T.T. Osteolysis: Basic science. *Clin. Orthop. Relat. Res.* **2001**, *393*, 71–77. [CrossRef]
21. Dorr, L.D.; Bloebaum, R.; Emmanual, J.; Meldrum, R. Histologic, biochemical, and ion analysis of tissue and fluids retrieved during total hip-arthroplasty. *Clin. Orthop. Relat. Res.* **1990**, *261*, 82–95. [CrossRef]
22. Jacobs, J.J.; Urban, R.M.; Hallab, N.J.; Skipor, A.K.; Fischer, A.; Wimmer, M.A. Metal-on-metal bearing surfaces. *J. Am. Acad. Orthop. Surg.* **2009**, *17*, 69–76. [CrossRef] [PubMed]
23. Ponthiaux, P.; Wenger, F.; Drees, D.; Celis, J.P. Electrochemical techniques for studying tribocorrosion processes. *Wear* **2004**, *256*, 459–468. [CrossRef]

24. Yan, Y.; Yang, H.; Su, Y.; Qiao, L. Study of the tribocorrosion behaviors of albumin on a cobalt-based alloy using scanning Kelvin probe force microscopy and atomic force microscopy. *Electrochem. Commun.* **2016**, *64*, 61–64. [CrossRef]

25. Uesaka, S.; Miyazaki, K.; Ito, H. Age-related changes and sex differences in chondroitin sulfate isomers and hyaluronic acid in normal synovial fluid. *Mod. Rheumatol.* **2004**, *14*, 470–475. [CrossRef] [PubMed]

26. Igual-Muñoz, A.; Mischler, S. Effect of the environment on wear ranking and corrosion of biomedical CoCrMo alloys. *J. Mater. Sci. Mater. Med.* **2011**, *22*, 437–450. [CrossRef] [PubMed]

27. Lewis, A.C.; Kilburn, M.R.; Heard, P.J.; Scott, T.B.; Hallam, K.R.; Allen, G.C.; Learmonth, I.D. The Entrapment of Corrosion Products from CoCr Implant Alloys in the Deposits of Calcium Phosphate: A Comparison of Serum, Synovial Fluid, Albumin, EDTA, and Water. *J. Orthop. Res.* **2006**, *24*, 1587–1596. [CrossRef] [PubMed]

28. Man, K.; Jiang, L.-H.; Foster, R.; Yang, X.B. Immunological response to total hip arthroplasty. *J. Funct. Biomater.* **2017**, *8*, 33. [CrossRef] [PubMed]

29. Lozano, R.M.; Pérez-Maceda, B.T.; Carboneras, M.; Onofre-Bustamante, E.; García-Alonso, M.C.; Escudero, M.L. Response of MC3T3-E1 osteoblasts, L929 fibroblasts and J774 macrophages to fluoride surface-modified AZ31 magnesium alloy. *J. Biomed. Mater. Res. Part A* **2013**, *101*, 2753–2762. [CrossRef] [PubMed]

30. VanOs, R.; Lildhar, L.L.; Lehoux, E.A.; Beaulé, P.E.; Catelas, I. In vitro macrophage response to nanometer-size chromium oxide particles. *J. Biomed. Mater. Res. Part B* **2014**, *102B*, 149–159. [CrossRef] [PubMed]

31. Billi, F.; Benya, P.; Kavanaugh, A.; Adams, J.; McKellop, H.; Ebramzadeh, E. The John Charnley Award: An accurate and extremely sensitive method to separate, display, and characterize wear debris: Part 2: Metal and ceramic particles. *Clin. Orthop. Relat. Res.* **2012**, *470*, 339–350. [CrossRef] [PubMed]

32. Kavanaugh, A.E.; Benya, P.; Billi, F. A Method to Isolate and Characterize Wear Debris from Synovial Fluid and Tissues. In *Metal-On-Metal Total Hip Replacement Devices*; ASTM International: West Conshohocken, PA, USA, 2013. [CrossRef]

33. Wang, J.; Witte, F.; Xi, T.; Zheng, Y.; Yang, K.; Yang, Y.; Zhao, D.; Meng, J.; Li, Y.; Li, W.; et al. Recommendation for modifying current cytotoxicity testing standards for biodegradable magnesium-based materials. *Acta Biomater.* **2015**, *21*, 237–249. [CrossRef] [PubMed]

34. Kruskal, W.H.; Wallis, W.A. Use of Ranks in One-Criterion Variance Analysis. *J. Am. Stat. Assoc.* **1952**, *47*, 583–621. [CrossRef]

35. Mann, H.B.; Whitney, D.R. On a Test of Whether one of Two Random Variables is Stochastically Larger than the Other. *Ann. Math. Stat.* **1947**, *18*, 50–60. [CrossRef]

36. Minitab Inc. Minitab Statistical Software. (n.d.). Available online: www.minitab.com/en-us/ (accessed on 16 April 2018).

37. R Core Team. *R: A Language and Environment for Statistical Computing*; R Foundation for Statistical Computing: Vienna, Austria, 2017; Available online: https://www.R-project.org/ (accessed on 16 April 2018).

Article

Experimentally Achievable Accuracy Using a Digital Image Correlation Technique in measuring Small-Magnitude (<0.1%) Homogeneous Strain Fields

Alice Acciaioli, Giacomo Lionello and Massimiliano Baleani *

IRCCS—Istituto Ortopedico Rizzoli, Laboratorio di Tecnologia Medica, 40136 Bologna, Italy;
alice.acciaioli@ior.it (A.A.); giacomo.lionello@ior.it (G.L.)
* Correspondence: baleani@tecno.ior.it; Tel.: +39-051-636-6865

Received: 20 March 2018; Accepted: 4 May 2018; Published: 8 May 2018

Abstract: Measuring small-magnitude strain fields using a digital image correlation (DIC) technique is challenging, due to the noise-signal ratio in strain maps. Here, we determined the level of accuracy achievable in measuring small-magnitude (<0.1%) homogeneous strain fields. We investigated different sets of parameters for image processing and imaging pre-selection, based on single-image noise level. The trueness of DIC was assessed by comparison of Young's modulus (E) and Poisson's ratio (ν) with values obtained from strain gauge measurements. Repeatability was improved, on average, by 20–25% with experimentally-determined optimal parameters and image pre-selection. Despite this, the intra- and inter-specimen repeatability of strain gauge measurements was 5 and 2.5 times better than DIC, respectively. Moreover, although trueness was also improved, on average, by 30–45%, DIC consistently overestimated the two material parameters by 1.8% and 3.2% for E and ν, respectively. DIC is a suitable option to measure small-magnitude homogeneous strain fields, bearing in mind the limitations in achievable accuracy.

Keywords: digital image correlation; homogeneous strain; small deformation level; accuracy; precision; calcium phosphate cements

1. Introduction

Calcium phosphate cements (CPCs) are bone substitute materials used for tissue defects filling [1]. Although they should mimic the mechanical behavior of bone tissue, their mechanical properties are still far from optimal. In fact, CPCs are brittle [2–4], and the limited data available in the literature suggests that this material can only withstand small strain levels (range 0.1–0.2%) before failure [5,6]. Therefore, new CPC formulations are still under development [7].

CPC testing is performed on small specimens (typically up to 20 mm in their largest dimension) [8–10]. Accurately measuring strain values the material undergoes to during testing is useful to compare different formulations of CPCs regarding elastic response and toughness enhancement. However, due to CPC characteristics, this is a challenge. In fact, contact-type extensometers cannot be used, because the knife edges would damage the specimen surface. Conversely, strain gauge installation may affect CPC response because of the unavoidable penetration of cyanoacrylate adhesive into the pores, which are always present in the cement matrix [8,11,12]. Therefore, the alternatives are non-contact techniques, i.e., optical methods based on interferometric techniques, or digital image correlation (DIC).

Interferometric techniques may be highly accurate, but are not practical for large sample size studies. They also may be too sensitive to environmental conditions, or could be limited to measurements on quasi-static loading scenarios [13–18]. Conversely, the DIC technique can be easily implemented, as demonstrated by its wide range of applications [19–23]. Video extensometry, initially based on feature-based image registration techniques that is evolving into a real-time DIC [24], can measure high strain level with small errors. However, strain errors increases if the specimen's dimensions decrease [25]. Moving on to DIC, experimental studies have defined effective procedures to calibrate DIC systems and/or to process the acquired images [26–29], thereby improving measurement accuracy. Unfortunately, although DIC has been demonstrated to be accurate in measuring medium-large strain levels [30–36], non negligible errors have been found when measuring small (<0.1% or 1000 microstrain) strain values [37–40], i.e., of the same order of magnitude of strain that CPCs can withstand before fracture. To the authors' knowledge, there is only one report showing that the DIC technique has the potential to provide an accuracy level comparable to the strain gauge in measuring small strain values [41]. However, that study was carried out analyzing artificial images, i.e., in absence of experimental errors. Conversely, studies based on experimentally-acquired images suggest that the measurement of small strain values may be affected by errors with an order of magnitude of up to 10% [28,38,39,42–44].

The present study investigated the suitability of DIC for measuring small-magnitude, homogeneous strain field, by experimentally determining the accuracy achievable using DIC to measure Young's modulus (E) and the Poisson's ratio (ν) of a specimen subjected to strain levels smaller than 0.1%.

2. Materials and Methods

The strain gauge (SG) technique was chosen as a reference technique to determine the error in measuring E and ν using the DIC technique. Aluminum was used instead of CPCs, to avoid any problems related to the SG application (described in the introduction section), i.e., an undesired reinforcing effect on the porous brittle material. Five parallelepiped specimens (10 mm square cross section, 20 mm height) were machined from an AA1050 (maximum grain size 80 μm) square bar, in order to ensure material uniformity. Each specimen underwent two series of uniaxial compression tests.

In the first series, a DIC system (Aramis 5M, GOM mbH, Braunschweig, Germany) was used to measure surface strain fields. Two digital cameras (2050 × 2448 pixels, TXG50i, Baumer Optronic GmbH, Radeberg, Germany), equipped with 2.8 FL/50 mm Titanar lenses and polarization filters (Schneider-Kreuznach, Bad Kreuznach, Germany) simultaneously monitored two opposite sides of the specimen at 15 Hz (Figure 1a).

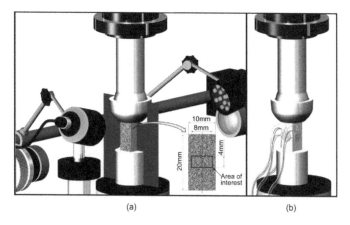

(a)   (b)

Figure 1. Scheme of the experimental setup used to measure surface strain. (**a**) Arrangement of the DIC system. The area of interest on the specimen surface is also shown. (**b**) Specimen instrumented with triaxial rosettes.

Both cameras were mounted onto a rigid support integrated into the testing machine frame and oriented in order to have the specimen surface centered in the view, while assuring a perpendicularity error smaller than 0.5 degrees. The camera-specimen distance was adjusted to the minimum focusing distance—about 150 mm from the polarizing filter placed in front of the camera lens to the specimen surface—obtaining a 12.6 × 15.0 mm measuring window. The resulting pixel size was 0.006 mm. A smaller pixel size, experimentally achievable with higher focal length lens, would have required a smaller dot size—determining the need to replace the airbrushing procedure (see below)—and would have reduced the subset dimension into millimeters (see below), thus introducing potential noise, due to the local strain gradient that occurs across grain boundaries. Speckle patterns (black dots on a white background) were previously created on two opposite surfaces of the specimen using an airbrush (Iwata HP-CH, 0.3 mm nozzle, Anest Iwata Europe S.r.l., Torino, Italy). Airbrush settings (air pressure = 3 bar; paint reduction by volume = 40%; airflow at the nozzle = 2 screw turns; needle travel length = 3 screw turns; spraying distance about 15 cm) were chosen using an internal airbrush-specific algorithm [45], in order to achieve an optimal speckle size of 3–5 pixels [46,47]. A trained operator (A.A.) created all patterns trying to achieve a coverage factor falling within the range of 42–50%, in order to minimise strain noise [48]. Average speckle size and coverage factor, calculated using the technique proposed by Lecompte et al. [49] and Shih [50], were 4.3 pixels (mean value range: 4.1–4.5 pixels) and 49% (range 47–50%) respectively. Images were acquired under the best achievable experimental conditions, i.e., at the smallest lens aperture (1/16, to get a sufficient depth of field), using the maximum exposure time (56 ms, due to the frame rate set to 15 Hz), while maximizing uniform lighting conditions over the entire area of interest (a white-light led was placed at about 80 mm from the specimen surface, the minimum allowable distance without interference of the 9W-led body with the field of view while maintaining the lamp heat-sink above and behind the camera lens). 2D calibration of each camera was carried out before each test-session by using a cubic-shaped 15 × 12 mm panel, following the manufacturer's recommendations [28].

An 8 by 4 mm rectangular area in the center of the specimen surface (Figure 1a) was selected as the area of interest, to exclude a local effect due to end effects. Preliminarily, the interaction between the subset size, step size, and strain window was investigated to select optimal parameters for image processing. The theoretically-set optimal subset size was calculated as three times the sum of the mean of the speckle size and distance (calculated values fell in the range 25 × 25–28 × 28 pixels), rounded to the nearest multiple of ten. Therefore, a subset size of 30 × 30 pixels was chosen. The step size was set at 1, 3, 8 and 15 pixels.

Image processing was carried out using a dedicated software (Aramis V 6.3.0, GOM mbH, Braunschweig, Germany). Surface strain in a measurement point, i.e., the center of a subset, was calculated using a square grid containing the center points of N × N neighboring subsets [51,52], hereinafter referred to as the strain window. The 2D deformation gradient tensor was calculated solving the system of equation, based on a first-order shape function, by the least square method [53,54]. The strain window was increased, starting from 3 × 3 subset, i.e., a field of center points of 3 × 3 subset, and increasing in steps of 2 subsets, up the maximum dimension of 9 × 9. Since increasing the step size decreased the noise in zero-strain maps (Figure 2), and the processing time without losing independent data, a step size of 15 pixels was chosen, i.e., an overlap ratio between neighboring subsets of 50% [55,56]. Similarly, a 9 × 9 strain window dimension was selected, because the noise decreased in zero-strain maps by increasing the strain window (Figure 2) [57]. Therefore the first image processing was carried out using a subset size of 30 × 30 pixels, a 50% overlap ratio and a 9 × 9 strain window.

A second image processing was carried out using the same parameters, except for the subset size, which was set at 60 × 60 pixels. This size was experimentally determined starting from the previous (30 × 30 pixels), and increasing it in steps of 10 pixels. The selection of the experimentally-determined optimal subset size to be used for image processing was based on the values of the coefficient of variation (CV, i.e., the ratio of the standard deviation to the mean, expressed in percent) of the principal strain values calculated for pooled data obtained from each couple of images. Indeed, CV values, which should theoretically be zero, are affected by both experimental and processing errors; the latter decreases by increasing the subset size. Since no further significant reduction was found when passing from pixel sizes 60×60 to 70×70, the 60×60 dimension was selected as the experimentally determined optimal subset size, as mentioned above (Figure 3, see also Supplementary Materials, Figure S1).

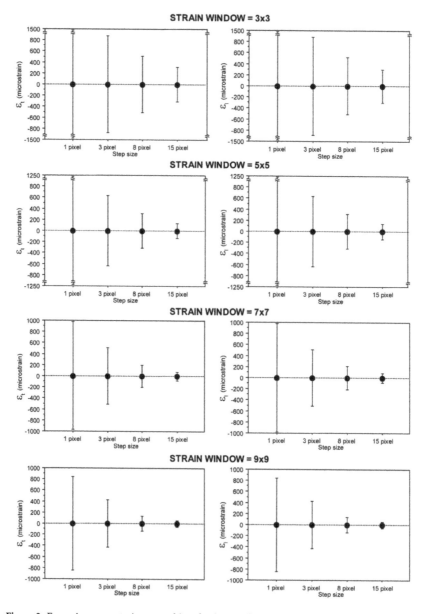

Figure 2. Errors in a zero-strain map achieved using a subset size of 30×30 pixels with different overlap values and strain window dimensions (error bar = standard deviation).

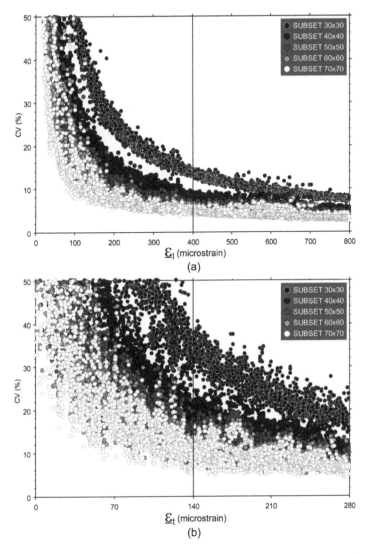

Figure 3. Representation of CV values calculated on the same image datasets changing the subset size from 30 × 30 to 70 × 70 pixels. (**a**) ε_l values; (**b**) ε_t values. Note: Each circle represents the CV of all strain values (ε) measured on the area of interest of a couple of images simultaneously acquired on both sides of the specimen.

A third image processing step was carried out using the previous parameters and an image pre-selection. Since noise on the strain map determines angle fluctuation of principal compressive (or longitudinal strain ε_l) and tensile strain (or transverse strain ε_t) direction, image pre-selection was based on the distribution of angle fluctuation: an image, and the image simultaneously acquired on the opposite surface of the specimen, was automatically discarded when three times the standard deviation of θ values ($3SD_\theta$) was greater than 0.1 rad, where θ was the angle between local ε_t and the horizontal direction (Figure 4).

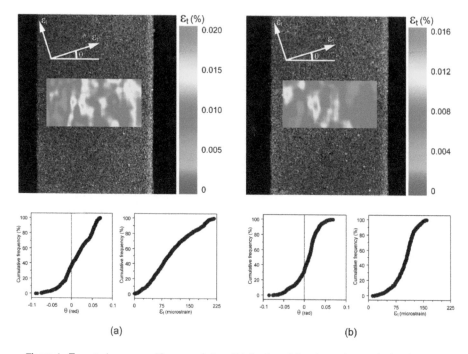

Figure 4. Two strain maps, with a cumulative distribution of θ and ε_t values, calculated for the same ε_l value (ε_l = 280 microstrain) on the same specimen surface during two different test-sessions. (a) $3SD_\theta$ > 0.1 rad: this image, and the image simultaneously acquired on the opposite surface of the specimen, must be discarded; (b) $3SD_\theta$ ≤ 0.1 rad: this image is accepted if θ distribution determined for the image simultaneously acquired on the opposite surface of the specimen fulfils the same requirement.

For all the three image processing, the average values (indicated as $\underline{\varepsilon}$) of ε_l and ε_t were calculated over the two areas of interest which were simultaneously acquired by the two cameras.

In the second series, triaxial rosettes with pre-attached lead wire (UFRA-3-350, Tokyo Sokki Kenkyuio Co, Tokyo, Japan) were used to measure strain reference values; this was necessary to determine the accuracy of the DIC technique (Figure 1b). After having carefully prepared the surface (Vishay Precision Group, 2014), a rosette was attached in the center of all four sides of the specimen surface (Figure 1b). Strain data were acquired at 100 Hz, using a multichannel data logger (System 6000, Vishay Precision Group, Raileigh, NC, USA), and processed to determine the principal directions and principal strains using a dedicated software (StrainSmart V4.31, Vishay Precision Group, Raileigh, NC, USA). The average values ($\underline{\varepsilon}$) of ε_l and ε_t simultaneously acquired by the four rosettes were calculated.

Both test series were performed at a constant displacement rate of 0.1 mm/min. Specimens were mounted onto a fixed platen placed onto a 10 kN load cell and loaded through an unlocked spherical seat platen fixed to the actuator of the testing machine (Mod.8502, Instron, Norwood, MA, USA), allowing alignment to the specimen end surface. A customized jig was used in order to align the specimen to the spherical seat, while maintaining the two patterned surfaces orthogonal to the DIC cameras. An initial preload (about 20 N) was applied to maintain the position of the specimen before removing the customized jig. All tests were limited to −5.6 kN, in order to reach a maximum compressive strain value just above a target value of 800 microstrain (i.e., 0.08%). The minimum compressive strain value was chosen considering the image processing errors. Indeed, referring to the experimentally-determined optimal subset size 60 × 60, CV values started to increase exponentially for ε_t values falling below 140 microstrain (Figure 3). Therefore this value, which corresponds to 400 microstrain in term of ε_l, was set as minimum strain value in calculating E and v.

Each test-session of five specimens was repeated five times after dismounting and remounting the whole setup. For each test repetition, E and v were calculated using ε_l and ε_t, measured as described above using both techniques. ε_l values were limited to a range of 400–800 microstrain for the aforementioned reasons. The E value was determined as the slope of the stress-ε_l curve. The v value was calculated as the mean value of all the absolute ratios of ε_t to ε_l. Inter- and intra-specimen repeatability was expressed as CV, calculated for five repetitions carried out on the same specimen and for each set of five specimens respectively. DIC trueness was assessed by analyzing differences between E and v mean values, which were determined using DIC and SG techniques on each specimen, by means of a paired t-test.

Finally, on the basis of the achieved results (see below), the third image processing was used to determine the elastic properties of a CPC prototype formulation. The formulation was optimized for 3D plotting, i.e., the paste allowed extrusion from a thin needle and assured printed shape stability. Ten parallelepiped specimens (10 mm square cross section, 20 mm height) were printed in air and immersed in deionized water for two weeks to achieve full setting. CPC specimens underwent monotonic compressive tests performed at a constant displacement rate of 0.1 mm/min. E and v were calculated limiting ε_l to the range of 400–800 microstrain for the aforementioned reasons.

3. Results

Noise in zero strain readings was negligible for SG; in all cases, measured values were smaller than 4 microstrain. Conversely, although ε_t and ε_l values calculated using DIC technique were smaller than 10 microstrain, ε standard deviation was 10–20 microstrain, with peak values up to 70 microstrain.

Approximately, 360 images were captured by each camera during a test against about 2400 triplets of data acquired from each rosette. Image pre-selection was used in the third processing procedure for DIC data. This process discarded, on average, 12% of images (range 0–35%). Discarded images were generally acquired in the first part of the test, i.e., where ε_t values calculated using DIC fell into the lower part of ε_t range. An example of the effect of image pre-selection on collected data is show in Figures 5 and 6.

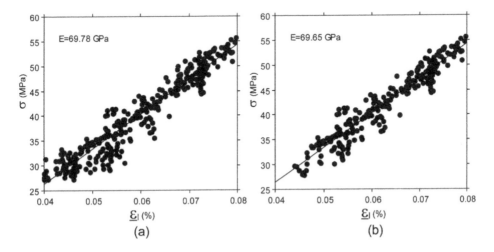

Figure 5. E value determined for one specimen. (**a**) Regression slope obtained using a subset size of 60 × 60 pixels. (**b**) Regression slope obtained using a subset size of 60 × 60 pixels and image pre-selection.

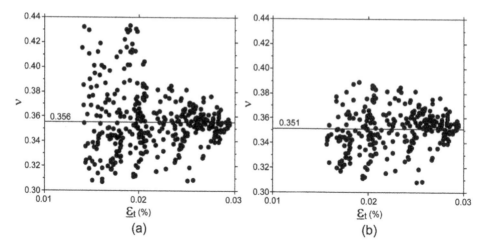

Figure 6. ν mean value determined for one specimen. (**a**) Data obtained using a subset size of 60×60 pixels. (**b**) Data obtained using a subset size of 60×60 pixels and image pre-selection. Note 1: The 140 microstrain threshold to ε_t had already been applied before calculating ν values reported in this graph. Note 2: Interchanging ε_l to ε_t in x-axis would: (i) scale the x-axis by a factor of $1/0.35$ (ii) cause a small horizontal drift of each point, the entity and direction of which would depend on the difference between the current value and 0.35; no changes would occur in ν mean values.

Intra- and inter-specimen repeatability values, calculated for the two techniques, are summarized in Tables 1 and 2 for E and ν values respectively. The three CV values reported for DIC technique refer to the three different image processing procedures. Intra- and inter-specimen repeatability of DIC were comparable. Conversely, inter-specimen repeatability of SG was double that of intra-specimen. In general, both techniques were more precise for determining E values. However, regardless of the measured parameter, the intra- and inter-specimen repeatability of SG measurements was better than DIC by about 5 and 2.5 times respectively. An improvement in DIC repeatability was found by increasing the subset size, and by adding image pre-selection. This improvement was more noticeable when looking at the worst CV values calculated for ν values.

Table 1. Intra- and inter-specimen repeatability for E values. The mean and the worst (maximum) CV values are reported.

Young's Modulus (E)	Strain Gauge Mean (Worst)	DIC Subset 30 × 30 Mean (Worst)	DIC Subset 60 × 60 Mean (Worst)	DIC Subset 60 × 60 + Image pre-Selec. Mean (Worst)
Intra-specimen repeatability	0.4% (0.7%)	2.2% (2.4%)	1.9% (2.2%)	1.8% (2.0%)
Inter-specimen repeatability	0.8% (1.0%)	2.3% (2.9%)	2.0% (2.7%)	1.9% (2.4%)

Table 2. Intra- and inter-specimen repeatability for ν values. The mean and the worst (maximum) CV values are reported.

Poisson's Ratio (ν)	Strain Gauge Mean (Worst)	DIC Subset 30 × 30 Mean (Worst)	DIC Subset 60 × 60 Mean (Worst)	DIC Subset 60 × 60 + Image Pre-Selec. Mean (Worst)
Intra-specimen repeatability	0.8% (1.2%)	4.4% (6.4%)	3.8% (5.4%)	3.3% (4.4%)
Inter-specimen repeatability	1.6% (2.2%)	4.3% (6.3%)	3.8% (5.5%)	3.5% (4.9%)

The trueness of DIC is shown in Figure 7, where values experimentally determined using SGs were used as reference. Although the bias decreased by increasing the subset size and adding image pre-selection, the difference was statistically significant in all three comparisons.

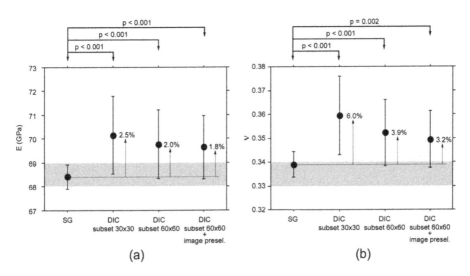

Figure 7. (a) E and (b) ν mean values, and relative standard deviations, determined using SG and DIC technique. The grey bands represent the range of E and ν values for AA1050, as reported in the literature. Note: The total sample size is 25, with 5 repeated measurements taken on the same specimen.

The pilot study on a CPC prototype formulation was successfully carried-out using the described experimental procedure. Indeed, elastic mechanical properties were determined for all ten tested specimens. E and ν were 9.0 ± 1.2 GPa and 0.22 ± 0.04 respectively.

4. Discussion

This study shows that DIC can measure small-magnitude (<0.1%) homogeneous strain fields with satisfactory accuracy, as demonstrated by the good agreement between E and ν values determined using SG (reference values) and DIC technique.

Although the DIC technique is extensively used to investigate material response to loads, these results were not foreseeable because, in general, published studies measured surface strain values with magnitudes of 1% or greater, both at macro and microscopic scales, [23,25,35,36,42,58–60] and reported error values of the same order of magnitude of the strain range as those investigated in the present study [28,39,42–44,61]. Indeed, the above-mentioned accuracy was achieved by minimizing the noise level, by (i) measuring each strain value (ε) on a 60×60 pixels subset, using a 50% overlap ratio (step size 30 pixel) and a 9×9 strain window, meaning that each strain value was calculated on a squared area of 300×300 pixels, i.e., 1.8×1.8 mm, and (ii) averaging all the strain values measured over an area of 64 mm² to calculate the ε_l and ε_t value for each couple of images simultaneously acquired, i.e., averaging about 900 ε values. Despite all this, some level of noise in ε_l and ε_t values, of up to 40 microstrain (in agreement with previous findings [62]), still remains, as shown by the dispersion of ν values in Figure 6. It has been widely demonstrated that the overlap ratio (or step size), subset size, and strain window dimension (or computation size) in image processing impact the DIC accuracy. These parameters were optimized in the present study for the described setup. Indeed, the chosen overlap ratio fell into the optimal range recommended in the literature, i.e., 50–75% [52,63,64]. A greater overlap ratio, i.e., smaller step size, would produce oversampled data, and increase noise on the strain field [48,65]. Similarly, a subset size greater than 60×60 pixels would not further reduce the residual error. It is acknowledged that increasing the subset size will increase the accuracy with which displacements are determined [49,66,67]. Indeed, CVs of ε values decrease as the subset size increases from 30×30 to 60×60, as shown in Figure 3. However, beyond a 60×60 pixel size, no further improvement was found; this is in agreement with previous studies showing that there is a plateau in noise reduction [25,42,66,67]. It must be acknowledged that subset sizes used in the literature are generally smaller than 60 × 60 pixels [30,31,68,69]. However, those studies investigated strain levels greater than 1%. Indeed, trends observed in Figure 3 suggest that subset size smaller than 60 × 60 become appropriate for measuring strain levels greater than 0.08%, thus increasing the ability of DIC to capture strain gradients. However when strain levels smaller that 0.1% are investigated, the optimal subset size increases [41,70]. It is noteworthy that CV values do not converge to zero. It seems unlikely that this noise is due to strain localization across grain boundaries, since the subset dimension, in millimeters, was at least one order of magnitude greater than the grain dimension. In other words, we assumed that the aluminum surface deformation appeared homogeneous at subset scale level. Noise in DIC measurements contributes to a residual trend. However, the actual strain distribution was non-uniform over the cross section, due to undesired off-axis loading on the specimen. In fact, although specimen alignment was performed carefully, a misalignment error of up to 0.1 mm may have occurred, which contributes up to 6.5% to residual CV value. The DIC setup (i.e., two cameras monitoring simultaneously two opposite side of the specimen), as well as the four-rosette configuration, was chosen in order to compensate for the misalignment error, and to determine accurate E value. Referring to the strain window dimension, a large value (9 × 9) was used, i.e., each ε value was calculated over a quite large area, as explained above. This choice decreases the ability of DIC to capture strain gradients, and therefore to accurately monitor what happens in proximity of any nucleating crack, which is irrelevant in the homogeneous strain field. Greater strain window dimensions might further lightly smooth ε values, but would not change the $\underline{\varepsilon}$ value calculated over the entire area of interest, and ultimately, would not change the results.

Other known sources of error are due to inaccuracies in preparing the specimen surface (speckle pattern), or in setting up the experimental model (loading condition, lens alignment, lighting, etc.). Special attention was paid in the specimen preparation. The speckle pattern was created by spraying black dots on the specimen surface following a previously developed procedure [45]. This involved the use of an airbrush-specific algorithm, which allows a trained operator to achieve the desired dot size of 3–5 pixels, and a coverage factor closer to 50%. On the basis of the data available in the literature [42,47,48,65,71], the achieved pattern should minimize the correlation error. In addition, the experimental model was set up attempting to achieve the best operating conditions. Additionally, the experimental setup minimized the error of perpendicularity between camera and specimen surface, decreasing the out-of-plane motion. Indeed, residual errors in calculating Poisson's ratio determined using the DIC technique, are comparable to those reported in the literature using the DIC 3D setup [72], which is unaffected by out-of-plane motion, without the need to manipulate DIC data. Finally, another source of error was deleted by simultaneously monitoring two opposite sides, in order to compensate for the problem of a non-axial-load applied on the specimen due to small misalignment between specimen axis and spherical seat. However, other errors are operator dependent (e.g., focusing camera lenses, lighting homogeneously both specimen surfaces). Despite the fact that the same trained operator carried out all test-sessions, the repeatability in achieving the experimental optimal conditions for the investigated strain range appeared to be a problem. However, it must be acknowledged that other sources of errors, such as thermal noise or CCD sensor noise [73] may have affected strain measurements. Whatever the cause, when experimental "sub-optimal" conditions occurred, image pre-selection played a role in reducing error and slightly improving DIC accuracy. In fact, image pre-selection discarded several images in all but one test session. The image pre-selection criterion is based on a priori knowledge of strain directions. It deletes all and only images that are noisy, which is more effective than data filtering, slightly improving both DIC trueness and precision in determining E and ν values (although a residual bias and a quite large dispersion still exists). Dispersion values can be ascribed to specimen inhomogeneity only for a small part. Indeed, apart from the fact that specimens were machined from the same bar, in order to minimize material inhomogeneity due to processing route, the inter-specimen repeatability achieved using SG was excellent. It could be argued that the E and ν values determined using SG are not true values, and therefore, that it is not possible to conclude that DIC overestimates the two mechanical parameters. However, although SG data are affected by some degree of uncertainty, the E and ν values determined using SG are in agreement with those reported in the literature for AA1050. This observation indirectly supports the aforementioned conclusion.

Other known sources of error are the quality of DIC hardware [28,74,75], and the correlation algorithm [34,37,54,74,76,77]. However, it was out of the scope of this study to compare different hardware or algorithms. Therefore, one commercial system was used to carry out the experimental series; this is a major limitation of this work. In fact, it cannot be excluded that different hardware components or different sub-pixel registration algorithms might have further improved the accuracy and precision achieved in the present study. However, the described procedure appeared suitable to determine CPC elastic mechanical properties, whose accuracy is a key for the reliability of numerical models of CPC 3D structures.

5. Conclusions

DIC can be practical and effective in measuring small-magnitude (<0.1%) homogeneous strain fields. The experimentally-determined optimal parameters for image processing, including the identification of the optimal subset size for the actual speckle pattern, and image pre-selection based on strain direction, can minimize the noise level. Under these conditions, a trueness better than 2% and 4% in measuring E and ν can be achieved, keeping in mind that both parameters are overestimated. However, DIC repeatability must be taken into account when calculating the sample size.

Supplementary Materials: The following are available online at http://www.mdpi.com/1996-1944/11/5/751/s1, Figure S1: Errors in a zero-strain map achieved using a subset size of 60 × 60 pixels with different overlap values and strain window dimensions (error bar = standard deviation).

Author Contributions: M.B. conceived and designed the experiments; A.A. and G.L. performed the experiments; A.A. and G.L. analyzed the data; A.A. and M.B. wrote the paper, with the help of G.L.

Funding: This research was funded by the Italian Program of Donation for Research "5 × 1000".

Acknowledgments: The authors wish to thank Luigi Lena for the artworks and Lucia Mancini for her help in revising the manuscript.

Conflicts of Interest: The authors declare no conflict of interest.

References

1. Zhang, J.; Liu, W.; Schnitzler, V.; Tancret, F.; Bouler, J.-M. Calcium phosphate cements for bone substitution: Chemistry, handling and mechanical properties. *Acta Biomater.* **2014**, *10*, 1035–1049. [CrossRef] [PubMed]
2. Ajaxon, I.; Öhman, C.; Persson, C. Long-Term In Vitro Degradation of a High-Strength Brushite Cement in Water, PBS, and Serum Solution. *BioMed Res. Int.* **2015**, *2015*, 575079. [CrossRef] [PubMed]
3. Ambard, A.J.; Mueninghoff, L. Calcium phosphate cement: Review of mechanical and biological properties. *J. Prosthodont.* **2006**, *15*, 321–328. [CrossRef] [PubMed]
4. Chow, L. Next generation calcium phosphate-based biomaterials. *Dent. Mater. J.* **2009**, *28*, 1–10. [CrossRef] [PubMed]
5. Charrière, E.; Terrazzoni, S.; Pittet, C.; Mordasini, P.; Dutoit, M.; Lemaître, J.; Zysset, P. Mechanical characterization of brushite and hydroxyapatite cements. *Biomaterials* **2001**, *22*, 2937–2945. [CrossRef]
6. Ajaxon, I.; Acciaioli, A.; Lionello, G.; Ginebra, M.P.; Öhman-Mägi, C.; Baleani, M.; Persson, C. Elastic properties and strain-to-crack-initiation of calcium phosphate bone cements: Revelations of a high-resolution measurement technique. *J. Mech. Behav. Biomed. Mater.* **2017**, *74*, 428–437. [CrossRef] [PubMed]
7. Geffers, M.; Groll, J.; Gbureck, U. Reinforcement Strategies for Load-Bearing Calcium Phosphate Biocements. *Materials* **2015**, *8*, 2700–2717. [CrossRef]
8. Engstrand, J.; Persson, C.; Engqvist, H. The effect of composition on mechanical properties of brushite cements. *J. Mech. Behav. Biomed. Mater.* **2014**, *29*, 81–90. [CrossRef] [PubMed]

9. Ginebra, M.P.; Driessens, F.C.M.; Planell, J.A. Effect of the particle size on the micro and nanostructural features of a calcium phosphate cement: A kinetic analysis. *Biomaterials* **2004**, *25*, 3453–3462. [CrossRef] [PubMed]

10. Luo, J.; Ajaxon, I.; Ginebra, M.P.; Engqvist, H.; Persson, C. Compressive, diametral tensile and biaxial flexural strength of cutting-edge calcium phosphate cements. *J. Mech. Behav. Biomed. Mater.* **2016**, *60*, 617–627. [CrossRef] [PubMed]

11. Grover, L. In vitro ageing of brushite calcium phosphate cement. *Biomaterials* **2003**, *24*, 4133–4141. [CrossRef]

12. Hofmann, M.; Mohammed, A.; Perrie, Y.; Gbureck, U.; Barralet, J. High-strength resorbable brushite bone cement with controlled drug-releasing capabilities. *Acta Biomater.* **2009**, *5*, 43–49. [CrossRef] [PubMed]

13. Davidson, R.J. *A Comparison of Moirè Interferometry and Digital Image Correlation*; Air force Institute of Technology: Dayton, OH, USA, 2008.

14. Erne, O.K.; Chu, Y.E.; Mohr, M.; Miller, J.R.; Bottlang, M. Full-field strain measurement on cortical bone. In Proceedings of the 48th Annual Meeting of the Orthopaedic Research Society, Dallas, TX, USA, 10–13 February 2002.

15. Jacquot, P. Speckle Interferometry: A Review of the Principal Methods in Use for Experimental Mechanics Applications. *Strain* **2008**, *44*, 57–69. [CrossRef]

16. Ohlson, N.G. Classical optical interferometry for strain measurement. In *Imaging Methods for Novel Materials and Challenging Applications*; Jin, H., Sciammarella, C., Furlong, C., Yoshida, S., Eds.; Springer: New York, NY, USA, 2013; Volume 3, pp. 127–132.

17. Post, D.H.B. Moirè Interfe. In *Handbook of Experimental Solid Mechanics*; Sharpe, N.W., Jr., Ed.; Springer: New York, NY, USA, 2008; pp. 1–26.

18. Yang, L.; Xie, X.; Zhu, L.; Wu, S.; Wang, Y. Review of electronic speckle pattern interferometry (ESPI) for three dimensional displacement measurement. *Chin. J. Mech. Eng.* **2014**, *27*, 1–13. [CrossRef]

19. Chevalier, L.; Calloch, S.; Hild, F.; Marco, Y. Digital image correlation used to analyze the multiaxial behavior of rubber-like materials. *Eur. J. Mech.-A/Solids* **2001**, *20*, 169–187. [CrossRef]

20. Sánchez-Arévalo, F.M.; Pulos, G. Use of digital image correlation to determine the mechanical behavior of materials. *Mater. Charact.* **2008**, *59*, 1572–1579. [CrossRef]

21. Tarigopula, V.; Hopperstad, O.S.; Langseth, M.; Clausen, A.H.; Hild, F. A study of localisation in dual-phase high-strength steels under dynamic loading using digital image correlation and FE analysis. *Int. J. Solids Struct.* **2008**, *45*, 601–619. [CrossRef]

22. Zhou, J.; Liu, D.; Shao, L.; Wang, Z. Application of digital image correlation to measurement of packaging material mechanical properties. *Math. Prob. Eng.* **2013**, *2013*, 204875. [CrossRef]

23. Lionello, G.; Fognani, R.; Baleani, M.; Sudanese, A.; Toni, A. Suturing the myotendinous junction in total hip arthroplasty: A biomechanical comparison of different stitching techniques. *Clin. Biomech.* **2015**, *30*, 1077–1082. [CrossRef] [PubMed]

24. Wu, R.; Kong, C.; Li, K.; Zhang, D. Real-Time Digital Image Correlation for Dynamic Strain Measurement. *Exp. Mech.* **2016**, *56*, 1–11. [CrossRef]

25. Pan, B.; Tian, L. Advanced video extensometer for non-contact, real-time, high-accuracy strain measurement. *Opt. Express* **2016**, *24*, 19082–19093. [CrossRef] [PubMed]

26. Hack, E.; Lin, X.; Patterson, E.A.; Sebastian, C.M. A reference material for establishing uncertainties in full-field displacement measurements. *Meas. Sci. Technol.* **2015**, *26*, 75004. [CrossRef]

27. Wang, D.; Diazdelao, F.A.; Wang, W.; Lin, X.; Patterson, E.A.; Mottershead, J.E. Uncertainty quantification in DIC with Kriging regression. *Opt. Lasers Eng.* **2016**, *78*, 182–195. [CrossRef]

28. Lava, P.; Van Paepegem, W.; Coppieters, S.; De Baere, I.; Wang, Y.; Debruyne, D. Impact of lens distortions on strain measurements obtained with 2D digital image correlation. *Opt. Lasers Eng.* **2013**, *51*, 576–584. [CrossRef]

29. Becker, T.; Splitthof, K.; Siebert, T.; Kletting, P. Error estimations of 3D digital image correlation measurements. *SPIE Proc.* **2006**, *6341*, 63410F.

30. Grytten, F.; Daiyan, H.; Polanco-Loria, M.; Dumoulin, S. Use of digital image correlation to measure large-strain tensile properties of ductile thermoplastics. *Polym. Test.* **2009**, *28*, 653–660. [CrossRef]

31. Jerabek, M.; Major, Z.; Lang, R.W. Strain determination of polymeric materials using digital image correlation. *Polym. Test.* **2010**, *29*, 407–416. [CrossRef]

32. Lagattu, F.; Bridier, F.; Villechaise, P.; Brillaud, J. In-plane strain measurements on a microscopic scale by coupling digital image correlation and an in situ SEM technique. *Mater. Charact.* **2006**, *56*, 10–18. [CrossRef]

33. Pritchard, R.H.; Lava, P.; Debruyne, D.; Terentjev, E.M. Precise determination of the Poisson ratio in soft materials with 2D digital image correlation. *Soft Matter* **2013**, *9*, 6037. [CrossRef]

34. Wang, Y.H.; Jiang, J.H.; Wanintrudal, C.; Du, C.; Zhou, D.; Smith, L.M.; Yang, L.X. Whole field sheet-metal tensile test using digital image correlation. *Exp. Tech.* **2010**, *34*, 54–59. [CrossRef]

35. Fuentes, C.A.; Willekens, P.; Hendrikx, N.; Lemmens, B.; Claeys, J.; Croughs, J.; Dupont-Gillain, C.; Seveno, D.; Van Vuure, A. Microstructure and mechanical properties of hemp technical fibres for composite applications by micro computed tomography and digital image correlation. In Proceedings of the 17th European Conference on Composite Materials (ECCM 2016), Munich, Germany, 26–30 June 2016; pp. 26–30.

36. Berfield, T.A.; Patel, J.K.; Shimmin, R.G.; Braun, P.V.; Lambros, J.; Sottos, N.R. Micro- and Nanoscale Deformation Measurement of Surface and Internal Planes via Digital Image Correlation. *Exp. Mech.* **2007**, *47*, 51–62. [CrossRef]

37. Chu, T.C.; Ranson, W.F.; Sutton, M.A. Applications of digital-image-correlation techniques to experimental mechanics. *Exp. Mech.* **1985**, *25*, 232–244. [CrossRef]

38. Hung, P.-C.; Voloshin, A.S. In-plane strain measurement by digital image correlation. *J. Braz. Soc. Mech. Sci. Eng.* **2003**, *25*, 215–221. [CrossRef]

39. Périé, J.-N.; Calloch, S.; Cluzel, C.; Hild, F. Analysis of a multiaxial test on a C/C composite by using digital image correlation and a damage model. *Exp. Mech.* **2002**, *42*, 318–328. [CrossRef]

40. Siebert, T.; Becker, T.; Spiltthof, K.; Neumann, I.; Krupka, R. Error Estimations in Digital Image Correlation Technique. *Appl. Mech. Mater.* **2007**, *7–8*, 265–270. [CrossRef]

41. Lee, C.; Take, W.A.; Hoult, N.A. Optimum Accuracy of Two-Dimensional Strain Measurements Using Digital Image Correlation. *J. Comput. Civ. Eng.* **2012**, *26*, 795–803. [CrossRef]

42. Rajan, V.P.; Rossol, M.N.; Zok, F.W. Optimization of Digital Image Correlation for High-Resolution Strain Mapping of Ceramic Composites. *Exp. Mech.* **2012**, *52*, 1407–1421. [CrossRef]

43. Begonia, M.T.; Dallas, M.; Vizcarra, B.; Liu, Y.; Johnson, M.L.; Thiagarajan, G. Non-contact strain measurement in the mouse forearm loading model using digital image correlation (DIC). *Bone* **2015**, *81*, 593–601. [CrossRef] [PubMed]

44. Robert, L.; Nazaret, F.; Cutard, T.; Orteu, J.-J. Use of 3-D Digital Image Correlation to Characterize the Mechanical Behavior of a Fiber Reinforced Refractory Castable. *Exp. Mech.* **2007**, *47*, 761–773. [CrossRef]

45. Lionello, G.; Cristofolini, L. A practical approach to optimizing the preparation of speckle patterns for digital-image correlation. *Meas. Sci. Technol.* **2014**, *25*, 107001. [CrossRef]

46. Schreier, H.; Orteu, J.-J.; Sutton, M.A. *Image Correlation for Shape, Motion and Deformation Measurements*; Springer: New York, NY, USA, 2009.

47. Zhou, P. Subpixel displacement and deformation gradient measurement using digital image/speckle correlation (DISC). *Opt. Eng.* **2001**, *40*, 1613. [CrossRef]

48. Carter, J.L.W.; Uchic, M.D.; Mills, M.J. Impact of Speckle Pattern Parameters on DIC Strain Resolution Calculated from In-situ SEM Experiments. *Fract. Fatigue Fail. Damage Evol.* **2015**, *5*, 119–126.

49. Lecompte, D.; Sol, H.; Vantomme, J.; Habraken, A. Analysis of speckle patterns for deformation measurements by digital image correlation. *Proc. SPIE* **2006**, *6341*. [CrossRef]

50. Shih, F.Y. *Image Processing and Mathematical Morphology*; CRC Press: New York, NY, USA, 2009.

51. Wang, Y.; Lava, P.; Reu, P.; Debruyne, D. Theoretical analysis on the measurement errors of local 2D DIC: Part II assessment of strain errors of the local smoothing method-Approaching an answer to the overlap question. *Strain* **2016**, *52*, 129–147. [CrossRef]

52. Eriksen, R.; Berggreen, C.; Boyd, S.W.; Dulieu-Barton, J.M. Towards high velocity deformation characterisation of metals and composites using Digital Image Correlation. *EPJ Web Conf.* **2010**, *6*, 31013. [CrossRef]

53. Schreier, H.W.; Sutton, M.A. Systematic errors in Digital Image Correlation due to unermatched subset shape functions. *Exp. Mech.* **2002**, *42*, 303–310. [CrossRef]

54. Pan, B.; Asundi, A.; Xie, H.; Gao, J. Digital image correlation using iterative least squares and pointwise least squares for displacement field and strain field measurements. *Opt. Lasers Eng.* **2009**, *47*, 865–874. [CrossRef]

55. Gu, X.; Pierron, F. Towards the design of a new standard for composite stiffness identification. *Compos. Part A Appl. Sci. Manuf.* **2016**, *91*, 448–460. [CrossRef]

56. Rossi, M.; Lava, P.; Pierron, F.; Debruyne, D.; Sasso, M. Effect of DIC Spatial Resolution, Noise and Interpolation Error on Identification Results with the VFM. *Strain* **2015**, *51*, 206–222. [CrossRef]

57. Lava, P.; Cooreman, S.; Coppieters, S.; De Strycker, M.; Debruyne, D. Assessment of measuring errors in DIC using deformation fields generated by plastic FEA. *Opt. Lasers Eng.* **2009**, *47*, 747–753. [CrossRef]
58. Joo, S.-H.; Lee, J.K.; Koo, J.-M.; Lee, S.; Suh, D.-W.; Kim, H.S. Method for measuring nanoscale local strain in a dual phase steel using digital image correlation with nanodot patterns. *Scr. Mater.* **2013**, *68*, 245–248. [CrossRef]
59. Kang, J.; Ososkov, Y.; Embury, J.; Wilkinson, D. Digital image correlation studies for microscopic strain distribution and damage in dual phase steels. *Scr. Mater.* **2007**, *56*, 999–1002. [CrossRef]
60. Ravindran, S.; Koohbor, B.; Kidane, A. On the Meso-Macro Scale Deformation of Low Carbon Steel. In *Advancement of Optical Methods in Experimental Mechanics*; Jin, H., Ed.; The Society for Experimental Mechanics: Bethel, CT, USA, 2015; pp. 409–414.
61. Pan, B. Bias error reduction of digital image correlation using Gaussian pre-filtering. *Opt. Lasers Eng.* **2013**, *51*, 1161–1167. [CrossRef]
62. Fayolle, X.; Calloch, S.; Hild, F. Controlling testing machines with digital image correlation. *Exp. Tech.* **2007**, *31*, 57–63. [CrossRef]
63. Koohbor, B.; Ravindran, S.; Kidane, A. Experimental determination of Representative Volume Element (RVE) size in woven composites. *Opt. Lasers Eng.* **2017**, *90*, 59–71. [CrossRef]
64. Leprince, S.; Barbot, S.; Ayoub, F.; Avouac, J.-P. Automatic and Precise Orthorectification, Coregistration, and Subpixel Correlation of Satellite Images, Application to Ground Deformation Measurements. *IEEE Trans. Geosci. Remote Sens.* **2007**, *45*, 1529–1558. [CrossRef]
65. Gu, J.; Cooreman, S.; Smits, A.; Bossuyt, S.; Sol, H.; Lecompte, D.; Vantomme, J. Full-field optical measurement for material parameter identification with inverse methods. *WIT Trans. Built Environ.* **2006**, *85*. [CrossRef]
66. Pan, B.; Xie, H.; Wang, Z.; Qian, K.; Wang, Z. Study on subset size selection in digital image correlation for speckle patterns. *Top. Appl. Phys. Opt. Express* **2008**, *16*, 7037–7048. [CrossRef]
67. Yaofeng, S.; Pang, J.H.L. Study of optimal subset size in digital image correlation of speckle pattern images. *Opt. Lasers Eng.* **2007**, *45*, 967–974. [CrossRef]
68. Carriero, A.; Abela, L.; Pitsillides, A.A.; Shefelbine, S.J. Ex vivo determination of bone tissue strains for an in vivo mouse tibial loading model. *J. Biomech.* **2014**, *47*, 2490–2497. [CrossRef] [PubMed]
69. Zhu, F.; Bai, P.; Zhang, J.; Lei, D.; He, X. Measurement of true stress–strain curves and evolution of plastic zone of low carbon steel under uniaxial tension using digital image correlation. *Opt. Lasers Eng.* **2015**, *65*, 81–88. [CrossRef]
70. Li, G.; Mubashar Hassan, G.; Dyskin, A.; MacNish, C. Study of Natural Patterns on Digital Image Correlation Using Simulation Method. *Int. J. Comput. Electr. Autom. Control Inf. Eng.* **2015**, *9*, 414–421.
71. Wang, Y.Q.; Sutton, M.A.; Bruck, H.A.; Schreier, H.W. Quantitative Error Assessment in Pattern Matching: Effects of Intensity Pattern Noise, Interpolation, Strain and Image Contrast on Motion Measurements. *Strain* **2009**, *45*, 160–178. [CrossRef]
72. Wittevrongel, L.; Badaloni, M.; Balcaen, R.; Lava, P.; Debruyne, D. Evaluation of Methodologies for Compensation of Out of Plane Motions in a 2D Digital Image Correlation Setup. *Strain* **2015**, *51*, 357–369. [CrossRef]
73. Haddadi, H.; Belhabib, S. Use of rigid-body motion for the investigation and estimation of the measurement errors related to digital image correlation technique. *Opt. Lasers Eng.* **2008**, *46*, 185–196. [CrossRef]
74. Barranger, Y.; Doumalin, P.; Dupré, J.C.; Germaneau, A. Digital Image Correlation accuracy: Influence of kind of speckle and recording setup. *EPJ Web Conf.* **2010**, *6*, 31002. [CrossRef]
75. Crammond, G.; Boyd, S.W.; Dulieu-Barton, J.M. Speckle pattern quality assessment for digital image correlation. *Opt. Lasers Eng.* **2013**, *51*, 1368–1378. [CrossRef]
76. Jiang, Z.; Kemao, Q.; Miao, H.; Yang, J.; Tang, L. Path-independent digital image correlation with high accuracy, speed and robustness. *Opt. Lasers Eng.* **2015**, *65*, 93–102. [CrossRef]
77. Tong, W. An Evaluation of Digital Image Correlation Criteria for Strain Mapping Applications. *Strain* **2005**, *41*, 167–175. [CrossRef]

Article

Finite Element Simulations of Hard-On-Soft Hip Joint Prosthesis Accounting for Dynamic Loads Calculated from a Musculoskeletal Model during Walking

Alessandro Ruggiero [1],*, Massimiliano Merola [2] and Saverio Affatato [2]

[1] Department of Industrial Engineering, University of Salerno, Via Giovanni Paolo II, nr. 132, 84084 Fisciano, Italy
[2] Medical Technology Laboratory, IRCCS—Rizzoli Orthopaedic Institute, Via di Barbiano, 1/10, 40136 Bologna, Italy; massimiliano.merola@ior.it (M.M.); affatato@tecno.ior.it (S.A.)
* Correspondence: ruggiero@unisa.it; Tel.: +39-089-964-312

check for updates

Received: 20 March 2018; Accepted: 5 April 2018; Published: 9 April 2018

Abstract: The hip joint replacement is one of the most successful orthopedic surgical procedures although it involves challenges to overcome. The patient group undergoing total hip arthroplasty now includes younger and more active patients who require a broad range of motion and a longer service lifetime for the replacement joint. It is well known that wear tests have a long duration and they are very expensive, thus studying the effects of geometry, loading, or alignment perturbations may be performed by Finite Element Analysis. The aim of the study was to evaluate total deformation and stress intensity on ultra-high molecular weight polyethylene liner coupled with hard material head during one step. Moving toward in-silico wear assessment of implants, in the presented simulations we used a musculoskeletal multibody model of a human body giving the loading and relative kinematic of the investigated tribo-system during the gait. The analysis compared two frictional conditions -dry and wet and two geometrical cases- with and without radial clearance. The loads and rotations followed the variability of the gait cycle as well as stress/strain acting in the UHWMPE cup. The obtained results allowed collection of the complete stress/strain description of the polyethylene cup during the gait and calculation of the maximum contact pressure on the lateral edge of the insert. The tensional state resulted in being more influenced by the geometrical conditions in terms of radial clearance than by the variation of the friction coefficients due to lubrication phenomena.

Keywords: total hip arthroplasty; musculoskeletal multibody model; dynamic loading; finite element analysis; radial clearance; dry and wet friction

1. Introduction

Total Hip Replacement (THR) is the most successful application of biomaterials in the short term in order to alleviate pain, restore joints, and increase functional mobility in diseased traumatized articulations [1–3]. A major limiting factor to the service life of THRs remains the wear of the polyethylene acetabular cup [4,5]. Preclinical endurance testing has become a standard procedure to predict the mechanical performance of new devices during their development. Wear tests are performed on different materials and designs used in prosthetic implants [1,6–8] to obtain quality assessment and acquire further knowledge about the tribological processes of joint prostheses. The objective of these investigations is to find out the wear rate and its dependence on the test conditions. In order to obtain realistic results, a wear test should reproduce the in vivo working conditions on the artificial implants [9].

It is well known that wear tests that are close to the in vivo conditions have a long duration and elevate costs [1,10,11]. The wear simulation is run for several million cycles, considering that one million cycles corresponds to one year in vivo [12–14]. The running-in period encompasses approximately the first half a million cycles; the steady-state wear is assessed by measuring the wear after the running-in period [15]. Wear is evaluated either gravimetrically, determining the weight loss of the components, or by measuring the volume of the material that has been removed, e.g., wear pit dimensions [9,16]. Lubricant absorption or creep of the loaded components, especially for polyethylene, is considered. Wear measurements are normally done at lubricant change stops, e.g., every 500,000 cycles.

Finite Element Analysis (FEA) has been widely used in many areas of biomechanics and medical engineering [17]. The numerical modelling tool of finite element analysis has been widely applied to study the behavior of articular cartilage, joints, and bone structures under compressive and tensile stresses. Structural applications include the design and development of joint prosthesis and fracture fixation devices. FEA enables to investigate parameters and boundary conditions, which are not accessible experimentally nor analytically. It has been applied to orthopedic devices to gain a deep understanding of the behavior of the bone-implant system and to support the design and pre-clinical testing of new devices. Thus, computational wear simulation can be a valuable complement to wear tests, e.g., in predictions of the lifetime of prostheses evaluating the effects of geometry, loading, or alignment perturbations.

As the first attempt to apply FEA to the orthopedic field dates back to 1972 [18], there have been four decades of developing and improving this methodology, along with the increase of computational power. Nevertheless, to make in silico wear simulation meaningful, better wear models are needed through advanced study design and corroboration with in vitro testing. The most intricate aspect of such models is the loading conditions acting on the implants, as their knowledge is still not fully reached. Few hip replacements have been subjected to direct load measures from Bergmann et al. [19–21], which limits the result validity to the subjects of the study. A more general load analysis can be obtained by muscular skeletal models, as it has been done by the authors [22], to study knee implants, or by van der Ploeg et al. [23], who used the multibody results as input for a FEA studying the micro-motions of a femoral stem.

In this work, a musculoskeletal multibody model was used in order to estimate the loads acting on the hip joint during walking. A finite elements analysis was then conducted using these loads and kinematic inputs. The study reports the results in terms of total deformation and stress intensity of an acetabular polyethylene liner coupled with a femoral head of a hard (ideally rigid) material.

2. Materials and Methods

2.1. Gait Cycles and Loads

In this study, the authors used a musculoskeletal modelling software to estimate loads acting on the hip joint during level walking. The simulation was performed with AnyBody Modelling System™ (AMS) [24]. To calculate the joint forces, knowledge of kinematic data and ground reaction forces that are used as input is required. These data are collected in gait analysis laboratories using special motion capture tools [25], which allows the measurement of a subject's gait through cameras that monitor markers on the subject's skin.

In inverse dynamics, the motion and the external loads on the body are known, and the aim is determining the internal forces. However, not enough equilibrium equations are available to find all the unknowns of the problem, therefore the calculation of the muscle forces is possible by the so-called redundancy problem. The solution of the muscle recruitment problem in the inverse dynamics approach is generally formulated as an optimization problem of the form:

$$\min G\left(f^{(M)}\right) \tag{1}$$

with

$$Cf = d \tag{2}$$

$$0 \leq f_i^{(M)} \leq N_i, i \in \{1, \ldots, n^{(M)}\}, \tag{3}$$

where G is the objective function (1), i.e., the assumed criterion of the recruitment strategy of the central nervous system, stated in terms of the muscle forces, and minimized with respect to all unknown forces in the problem, $f = \left(f^{(M)T} f^{(R)T} \right)^T$, (i.e., muscle forces and joint reactions). Equation (2) is the dynamic equilibrium equations, which enter as constraints into the optimization. C is the coefficient matrix for the unknown forces/moments in the system, d is a vector of the known applied loads and inertia forces. The non-negativity constraints on the muscle forces, Equation (3), states that muscle can only pull, not push and the upper bounds limit their capability, so N_1 is the strength of the muscle.

The most popular forms of the objective function G, calculated from the relative intensity and normalized for each muscle, are the polynomial criteria and the soft saturation criteria per Siemienski et al. [26]:

$$G(f^{(M)}) = \sum_{i=1}^{n^{(M)}} \left(\frac{f_i^{(M)}}{N_i} \right)^p \tag{4}$$

with $p = 2$, since it is established in the literature as predicting reasonable muscle activation patterns for the type of analyzed trial.

All segments of the biomechanical system are modelled as rigid bodies, neglecting effects such as the wobbly masses of soft tissues.

2.2. Finite Element Modelling

The finite elements model was realized through Ansys® Workbench commercial software (v.18.1, ANSYS Inc., Canonsburg, PA, USA). To minimize the computational complexity there were considered only the two major bearing components in a hip implant, namely the femoral head and the acetabular cup. The presence of the pelvic bone has been neglected, since its influence has a least effect on contact pressure [27]. The bodies of the FE model are shown in Figure 1, the mesh was realized through quadratic tetrahedral elements. Mesh convergence tests were performed, resulting in a total number of elements and nodes of 2012 and 3283, respectively, having 383 contact elements.

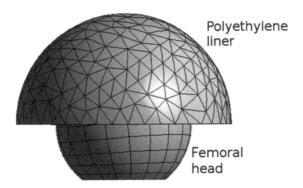

Figure 1. Mesh model of the femoral head and the polyethylene liner.

To solve the contact problem the Augmented Lagrange algorithm was used, considering an asymmetric behavior, which is recommended for the solution of frictional contacts. The femoral head was modelled as a rigid body, whereas for the polyethylene insert of the cup it was selected the Ultra-High-Molecular-Weight-Polyethylene (UHMWPE) GUR 1050 material, as it is one of the main polymer materials used for implant applications [28,29]; Table 1 summarizes its main parameters.

Table 1. UHMWPE GUR 1050 attributes.

Density (kg·m^{-3})	Young's Modulus (MPa)	Poisson's Ratio (-)	Bulk Modulus (MPa)	Shear Modulus (MPa)	Tensile Yield Strength (MPa)	Tensile Ultimate Strength (MPa)
930	690	0.43	1640	241	21	40

The material is assumed to be homogenous and isotropic. Even if roughness plays a key role on the tribological behavior of hip implants [30,31], the model surfaces are considered smooth, as in most of the models found in literature, for the ease of solution.

The study was conducted to understand the stress behaviour in dry conditions, as the presence of lubricant modifies the pressure distribution on the surfaces. However, as further knowledge, the influence of the friction coefficient was introudced, comparing a dry and a lubricated regime, namely, a DRY case, i.e., no lubrication, and a WET case, i.e., lubricated conditions. The mean values of the friction coefficient in lubricated conditions were extracted from experimental studies on the prosthesis tribological pairs [32,33]. Thus, a dry friction value of 0.13 and a mean wet friction value of 0.05 was selected.

The femoral head had a diameter of 28 mm, whereas the acetabular cup had a thickness of 5 mm. Moreover, two geometrical configurations of the coupled bodies were studied, and the simulations were executed considering the presence of the radial clearance (CC condition) and not considering it (NC condition). Radial clearance is the difference between the radius of the acetabular cup and the one of the femoral head (see Figure 2). Radial clearance, when considered, was 0.5 mm [34]; the results of the two configurations were then compared.

Radial clearance = R$_c$ - R$_h$

Figure 2. Radial clearance is the difference in radius of the acetabular cup and the femoral head. In this image the clearance is amplified for a better understanding.

The input data obtained by the multibody system were evaluated in a local coordinates system, which follows the movements of the femoral head. As the FE model requires a global coordinates system, a conversion was performed considering well-known geometric transformations [35]. Further force components and mesh orientations are defined with respect to a pelvic reference frame that coincides with the true anatomic superior, anterior, and lateral directions (see Figure 3). The tensional state and the total deformation were evaluated on the inner surface of the polyethylene liner.

3. Results

The forces and the rotations taking place along the three degrees of freedom were gained from the multibody analysis. The force components are the ones depicted in Figure 3. The three rotations around the axes are the *Flexion/Extension* (around *z* axes), *Abduction/Adduction* (around *x* axes) and *Inward/Outward* (around *y* axes).

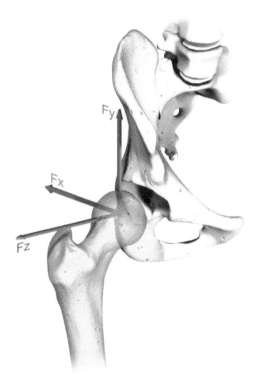

Figure 3. The three force vectors represented on a femoral head.

In Figure 4 the force components and the rotations derived from the model are shown. It is noticeable that the highest rotation is the *Flexion/Extension*, whereas the highest load is along the *y* axes. Forces and rotations so obtained were used as dynamic inputs for the finite element model.

In Figure 5 the pressure distribution on the internal surface of the cup in different instants of the cycle is shown, with regards to the dry NC condition as exemplificative case. Along with the different orientation of the femoral head it is possible to observe the pressure distribution on the polyethylene liner. Its highest values are found at 8% and 48% of the cycle (respectively Figure 5a,c). In the latter case the highest level of the *Anterior/Posterior* force was also found (see Figure 4), and the pressure is more concentrated in the edge zone of the insert. In the other two images, Figure 5b,d, at 26% and 93% of the cycle, the pressure reaches lower values and its mostly located in the central part of the inner hemisphere. The other geometrical and frictional cases are here omitted for brevity, but they presented a similar distribution of pressure, only leading different intensity of the tensional state.

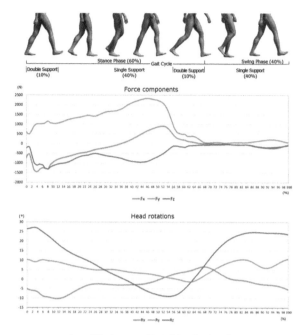

Figure 4. Forces components, in red Fx, in green Fy, in blue Fz. Head rotations, around *x* axes in red, *y* axes green and *z* axes blue.

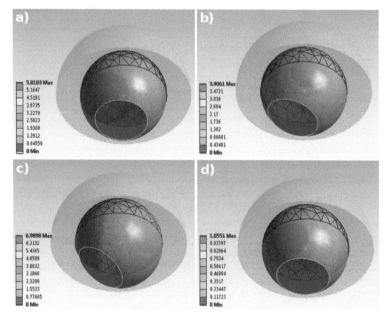

Figure 5. Pressure distribution on the internal surface of the acetabula cup. (**a**) NC case, dry condition; (**b**) NC case, wet condition; (**c**) CC case, dry condition; (**d**) CC case, wet condition.

Figures 6 and 7 present a summary of the maximum values of pressure and total deformation, comparing the two friction conditions and the two geometric configurations. In Figure 6a the maximum value of pressure is shown, for each part of the cycle. The maximum value throughout the walking cycle is equal to 7 MPa and it is found around 48% of the way through the cycle—agreeing with the dynamic analysis. A slight difference was found between the two friction cases, showing a higher peak value in the wet condition. In Figure 6c the maximum value of the total deformation is displayed; its highest value is again related to the wet case, reaching almost 0.6 mm. In Figure 6b,d the comparisons are shown, in terms of pressure and deformation, between the two friction cases considering the presence of the radial clearance (CC). In Figure 6b, the comparison highlights the almost complete lack of difference in the two friction conditions; the highest value reached is almost 10 MPa. However, in Figure 6d the curves have some differences, showing slightly higher values of total deformation in wet condition (maximum value of 1 mm and 0.9 mm for the wet and the dry case, respectively).

In Figure 7 the comparison of the two geometrical conditions is presented, with and without radial clearance (CC and NC, respectively). In Figure 7a, considering the boundary lubrication, the maximum pressure plot shows the large divergence in the two geometrical solutions. The highest values are almost 10 MPa and 6.7 MPa, for the CC and NC respectively. As well as the curves in Figure 7c, where the total deformation is shown, the highest values are found for the CC condition where it reaches almost 1 mm, whereas NC gives back at the most 0.6 mm. The curves in Figure 7b also brings out the divergences in the two geometrical configurations under dry friction, the pressure in CC being higher than the one found in NC; the values are almost the same as those already described in the wet condition. Figure 7d reports the maximum values of the total deformation, clearly showing the difference in the CC and NC conditions.

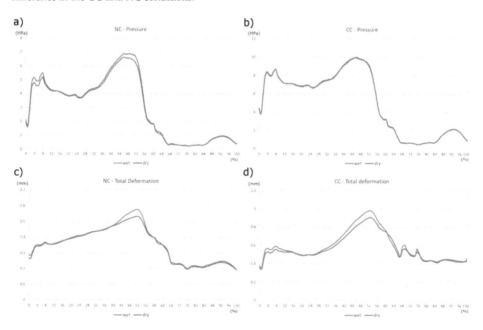

Figure 6. Comparison of result in the NC (**a,c**) and CC (**b,d**) cases, between the dry and wet conditions.

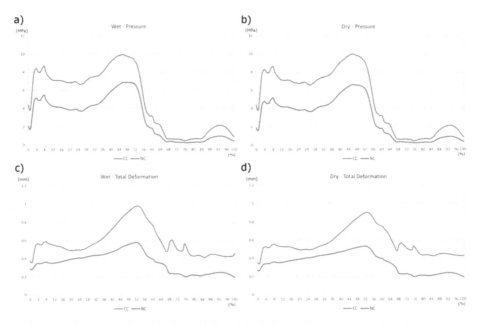

Figure 7. Comparison of result in the wet (**a,b**) and dry (**c,d**) condition, between the CC and NC cases.

4. Discussion

In this work, the forces and rotations acting on the hip joint during a complete gait cycle were first evaluated by solving the inverse dynamic problem. These data agree with the walking cycle variability [25], showing how the joint results more loaded during the stance phase than in the swing phase, when there is almost no contact between the foot and the ground. Furthermore, the highest loads is the axial (along the y axes) as expected and confirmed by a comparison with the standard ISO 14242-1:2012 [36]. Afterwards, these loads and rotation were used as dynamical input in the finite element model, obtaining the pressure distribution and the elastic deformation on each node of the surface during the whole gait cycle. The stress state of the internal surface of the UHMWPE liner reaches peaks of the order of 10 MPa, being in accordance with literature results [37–39]. Several authors [36,40,41] used simplified hip kinematic and gait load, such as one-dimensional vertical load or considered only the flexion/extension rotation [40], which does not represent actual physiological loading, plus these studies rely on ISO standards at the expense of a flexible design. In this regard, the original contribution of the study is the acquisition of stress and deformation distribution based on a multibody model. This offers the possibility to apply a wide variety of gaits and to consider the peculiarity of a specific class of patients, e.g., older or younger people. This could lead to the design of prosthesis which account the requirement of a patient, by simply characterize his activity through the multibody and finite elements model.

The comparison of the different working conditions allowed the conclusion that the presence of radial clearance strongly influences the tensional state of the coupled surfaces. Whereas, for a complete understanding of the lubrication influence on the tribology of hip implants the Reynolds equations must be involved [42], as varying the coefficient of friction does not allow the establishment of a sensible difference in results. In literature, the contact is usually assumed to be frictionless, justifying this hypothesis with negligible variation in contact pressure distribution. However, the presence of a frictional force has an influence to the location of the nominal contact point, as demonstrated by Mattei and Di Puccio in [35].

The reasons for the differences across the two geometrical conditions are ascribable to the different tensional state of the surfaces. In the geometry without radial clearance a conformal contact is realized, meaning that the contact is distributed along an area. On the other hand, when the radial clearance is considered, the contact become non-conformal, which implies that the contact is limited to a small area that increases its extension as the deformation rises. These differences in the area size led to the distinction of the pressure values, shown in the previous paragraph. As stated by Teoh et al. [37] the radial clearance plays a vital role in the wear process, finding that its extreme values (0.001 and 0.5 mm) led to the highest level of wear. In the study by Tudor et al. [43] the difference between the conditions of small loads with high clearances is highlighted, where the contact pressure tends to a Hertzian distribution, and high loads with small clearances, where the contact pressure tends to have a hyperbolic distribution. Furthermore, concerning the influence of the frictional force, in the study by Teoh et al. [37] they found that varying the friction coefficient between 0 and 0.3, only yielded a slightly variance of the wear rate.

5. Conclusions

This work is meant as a first step forward to the simplified hip kinematic and gait load. In this work we presented a dynamic load, derived by a multibody technique applied to a musculoskeletal model, considering the variability of the load direction during gait movement.

The main conclusions of the study are:

- multibody technique applied to a musculoskeletal model was proven to be a valid instrument to obtain a flexible design of implant, leading to the evaluation of load for a specific demanding task;
- load components and rotations match the variability between the stance and the swing phase of the leg during the gait;
- highest values of pressure on the inner surface of the polyethylene insert are found near its edge side;
- tensional state and elastic deformation are mainly influenced by the radial clearance rather than the friction coefficient.

The results of the present work offer the possibility to extend, in further studies, the range of kinematics and dynamics; an additional improvement could include lubrication to apply this set-up in a realistic scenario.

Acknowledgments: The authors would like to thank Simone Di Stasio (University of Salerno) for his help with the finite element model. This work was partially supported by the Italian Program of Donation for Research "5 per mille", year 2014, and by FARB 2016 obtained from University of Salerno.

Author Contributions: A.R. and S.A. conceived and designed the investigation; A.R. developed the dynamical analysis; M.M. performed the FEM model; M.M. and S.A. analyzed the data; A.R. and S.A. discussed the results; A.R., M.M. and S.A. wrote the paper.

Conflicts of Interest: The authors declare no conflict of interest.

Nomenclature

THR	Total Hip Replacement
AMS	AnyBody Modelling System
FEA	Finite Element Analysis
UHMWPE	Ultra-High-Molecular-Polyethylene
G	objective function
$f^{(M)}$	muscle forces
$f^{(R)}$	joint reaction
d	vector of applied loads and inertia forces
N_1	strength of the muscle

References

1. Affatato, S.; Freccero, N.; Taddei, P. The biomaterials challenge: A comparison of polyethylene wear using a hip joint simulator. *J. Mech. Behav. Biomed. Mater.* **2016**, *53*, 40–48. [CrossRef] [PubMed]
2. Learmonth, I.D.; Young, C.; Rorabeck, C. The operation of the century: Total hip replacement. *Lancet* **2007**, *370*, 1508–1519. [CrossRef]
3. Kurtz, S.; Ong, K.; Lau, E.; Mowat, F.; Halpern, M. Projections of Primary and Revision Hip and Knee Arthroplasty in the United States from 2005 to 2030. *J. Bone Jt. Surg.* **2007**, *89*, 780–785. [CrossRef]
4. Affatato, S.; Ruggiero, A.; Merola, M. Advanced biomaterials in hip joint arthroplasty. A review on polymer and ceramics composites as alternative bearings. *Compos. Part B Eng.* **2015**, *83*, 276–283. [CrossRef]
5. Rajaee, S.S.; Theriault, R.V.; Pevear, M.E.; Smith, E.L. National Trends in Primary Total Hip Arthroplasty in Extremely Young Patients: A Focus on Bearing Surface Usage From 2009 to 2012. *J. Arthroplast.* **2016**, *31*, 63–68. [CrossRef] [PubMed]
6. Affatato, S.; Bersaglia, G.; Emiliani, D.; Foltran, I.; Taddei, P.; Reggiani, M.; Ferrieri, P.; Toni, A. The performance of gamma- and EtO-sterilised UHMWPE acetabular cups tested under severe simulator conditions. Part 2: Wear particle characteristics with isolation protocols. *Biomaterials* **2003**, *24*, 4045–4055. [CrossRef]
7. Affatato, S.; Bersaglia, G.; Rocchi, M.; Taddei, P.; Fagnano, C.; Toni, A. Wear behaviour of cross-linked polyethylene assessed in vitro under severe conditions. *Biomaterials* **2005**, *26*, 3259–3267. [CrossRef] [PubMed]
8. Affatato, S.; Ruggiero, A.; Jaber, S.; Merola, M.; Bracco, P. Wear Behaviours and Oxidation Effects on Different UHMWPE Acetabular Cups Using a Hip Joint Simulator. *Materials* **2018**, *11*, 433. [CrossRef] [PubMed]
9. Grillini, L.; Affatato, S. How to measure wear following total hip arthroplasty. *Hip Int.* **2013**, *23*, 233–242. [CrossRef] [PubMed]
10. Affatato, S.; Zavalloni, M.; Spinelli, M.; Costa, L.; Bracco, P.; Viceconti, M. Long-term in-vitro wear performance of an innovative thermo-compressed cross-linked polyethylene. *Tribol. Int.* **2010**, *43*, 22–28. [CrossRef]
11. Affatato, S.; Testoni, M.; Cacciari, G.L.; Toni, A. Mixed oxides prosthetic ceramic ball heads. Part 2: Effect of the ZrO_2 fraction on the wear of ceramic on ceramic joints. *Biomaterials* **1999**, *20*, 971–975. [CrossRef]
12. Affatato, S.; Spinelli, M.; Zavalloni, M.; Mazzega-Fabbro, C.; Viceconti, M. Tribology and total hip joint replacement: Current concepts in mechanical simulation. *Med. Eng. Phys.* **2008**, *30*, 1305–1317. [CrossRef] [PubMed]
13. Dumbleton, J.H. *Tribology of Natural and Artificial Joints*; Elsevier: New York, NY, USA, 1981.
14. Petersen, D.; Link, R.; Wang, A.; Polineni, V.; Essner, A.; Sokol, M.; Sun, D.; Stark, C.; Dumbleton, J. The Significance of Nonlinear Motion in the Wear Screening of Orthopaedic Implant Materials. *J. Test. Eval.* **1997**, *25*, 239–245. [CrossRef]
15. Affatato, S.; Bersaglia, G.; Foltran, I.; Emiliani, D.; Traina, F.; Toni, A. The influence of implant position on the wear of alumina-on-alumina studied in a hip simulator. *Wear* **2004**, *256*, 400–405. [CrossRef]
16. Astarita, A.; Rubino, F.; Carlone, P.; Ruggiero, A.; Leone, C.; Genna, S.; Merola, M.; Squillace, A. On the Improvement of AA2024 Wear Properties through the Deposition of a Cold-Sprayed Titanium Coating. *Metals* **2016**, *6*, 185. [CrossRef]
17. Taylor, M.; Prendergast, P.J. Four decades of finite element analysis of orthopaedic devices: Where are we now and what are the opportunities? *J. Biomech.* **2015**, *48*, 767–778. [CrossRef] [PubMed]
18. Brekelmans, W.A.M.; Poort, H.W.; Slooff, T.J.J.H. A New Method to Analyse the Mechanical Behaviour of Skeletal Parts. *Acta Orthop. Scand.* **1972**, *43*, 301–317. [CrossRef] [PubMed]
19. Bergmann, G.; Graichen, F.; Rohlmann, A. Hip joint loading during walking and running, measured in two patients. *J. Biomech.* **1993**, *26*, 969–990. [CrossRef]
20. Bergmann, G.; Bergmann, G.; Deuretzabacher, G.; Deuretzabacher, G.; Heller, M.; Heller, M.; Graichen, F.; Graichen, F.; Rohlmann, A.; Rohlmann, A.; et al. Hip forces and gait patterns from routine activities. *J. Biomech.* **2001**, *34*, 859–871. [CrossRef]
21. Bergmann, G.; Kniggendorf, H.; Graichen, F.; Rohlmann, A. Influence of shoes and heel strike on the loading of the hip joint. *J. Biomech.* **1995**, *28*, 817–827. [CrossRef]
22. Ruggiero, A.; Merola, M.; Affatato, S. On the biotribology of total knee replacement: A new roughness measurements protocol on in vivo condyles considering the dynamic loading from musculoskeletal multibody model. *Meas. J. Int. Meas. Confed.* **2017**, *112*. [CrossRef]
23. Van der Ploeg, B.; Tarala, M.; Homminga, J.; Janssen, D.; Buma, P.; Verdonschot, N. Toward a more realistic prediction of peri-prosthetic micromotions. *J. Orthop. Res.* **2012**, *30*, 1147–1154. [CrossRef] [PubMed]

24. Damsgaard, M.; Rasmussen, J.; Christensen, S.T.; Surma, E.; de Zee, M. Analysis of musculoskeletal systems in the AnyBody Modeling System. *Simul. Model. Pract. Theory* **2006**, *14*, 1100–1111. [CrossRef]

25. Vaughan, C.; Davis, B.L.; O'Connor, J.C. The man data set from "Dynamics of Human Gait". *Hum. Kinet. Publ.* 1992. Available online: https://isbweb.org/data (accessed on 6 April 2018).

26. Siemienski, A. Soft saturation—An idea for load sharing between muscles. Application to the study of human locomotion. In *Biolocomotion: A Century of Research Using Moving Pictures*; Cappozzo, A., Marchetti, M., Tosi, V., Eds.; Promograph: Rome, Italy, 1992; pp. 293–303.

27. Barreto, S.; Folgado, J.; Fernandes, P.R.; Monteiro, J. The Influence of the Pelvic Bone on the Computational Results of the Acetabular Component of a Total Hip Prosthesis. *J. Biomech. Eng.* **2010**, *132*, 54503. [CrossRef] [PubMed]

28. Kurtz, S.M.; Villarraga, M.L.; Herr, M.P.; Bergström, J.S.; Rimnac, C.M.; Edidin, A.A. Thermomechanical behavior of virgin and highly crosslinked ultra-high molecular weight polyethylene used in total joint replacements. *Biomaterials* **2002**, *23*, 3681–3697. [CrossRef]

29. Laurian, T.; Tudor, A. Some Aspects Regarding the Influence of the Clearance on the Pressure Distribution in Total Hip Joint Prostheses. In Proceedings of the National Tribology Conference RotTrib03, Galati, Romania, 24–26 September 2003.

30. Merola, M.; Ruggiero, A.; de Mattia, J.S.; Affatato, S. On the tribological behavior of retrieved hip femoral heads affected by metallic debris. A comparative investigation by stylus and optical profilometer for a new roughness measurement protocol. *Measurement* **2016**, *90*, 365–371. [CrossRef]

31. Affatato, S.; Ruggiero, A.; Merola, M.; Logozzo, S. Does metal transfer differ on retrieved Biolox® Delta composites femoral heads? Surface investigation on three Biolox® generations from a biotribological point of view. *Compos. Part B Eng.* **2017**, *113*, 164–173. [CrossRef]

32. Ruggiero, A.; D'Amato, R.; Gómez, E. Experimental analysis of tribological behavior of UHMWPE against AISI420C and against TiAl6V4 alloy under dry and lubricated conditions. *Tribol. Int.* **2015**, *92*, 154–161. [CrossRef]

33. Ruggiero, A.; D'Amato, R.; Gómez, E.; Merola, M. Experimental comparison on tribological pairs UHMWPE/TIAL6V4 alloy, UHMWPE/AISI316L austenitic stainless and UHMWPE/AL$_2$O$_3$ ceramic, under dry and lubricated conditions. *Tribol. Int.* **2016**, *96*, 349–360. [CrossRef]

34. Shen, F.-W.; Lu, Z.; McKellop, H.A. Wear versus Thickness and Other Features of 5-Mrad Crosslinked UHMWPE Acetabular Liners. *Clin. Orthop. Relat. Res.* **2011**, *469*, 395–404. [CrossRef] [PubMed]

35. Mattei, L.; di Puccio, F. Wear Simulation of Metal-on-Metal Hip Replacements with Frictional Contact. *J. Tribol.* **2013**, *135*, 21402. [CrossRef]

36. Matsoukas, G.; Willing, R.; Kim, I.Y. Total Hip Wear Assessment: A Comparison Between Computational and In Vitro Wear Assessment Techniques Using ISO 14242 Loading and Kinematics. *J. Biomech. Eng.* **2009**, *131*, 41011. [CrossRef] [PubMed]

37. Teoh, S.H.; Chan, W.H.; Thampuran, R. An elasto-plastic finite element model for polyethylene wear in total hip arthroplasty. *J. Biomech.* **2002**, *35*, 323–330. [CrossRef]

38. Maxian, T.A.; Brown, T.D.; Pedersen, D.R.; Callaghan, J.J. A sliding-distance-coupled finite element formulation for polyethylene wear in total hip arthroplasty. *J. Biomech.* **1996**, *29*, 687–692. [CrossRef]

39. Gao, Y.; Jin, Z.; Wang, L.; Wang, M. Finite element analysis of sliding distance and contact mechanics of hip implant under dynamic walking conditions. *Proc. Inst. Mech. Eng. Part H J. Eng. Med.* **2015**, *229*, 469–474. [CrossRef] [PubMed]

40. Bevill, S.L.; Bevill, G.R.; Penmetsa, J.R.; Petrella, A.J.; Rullkoetter, P.J. Finite element simulation of early creep and wear in total hip arthroplasty. *J. Biomech.* **2005**, *38*, 2365–2374. [CrossRef] [PubMed]

41. Maxian, T.A.; Brown, T.D.; Pedersen, D.R.; Callaghan, J.J. Adaptiive finite element modeling of long-term polyethylene wear in total hip arthroplasty. *J. Orthop. Res.* **1996**, *14*, 668–675. [CrossRef] [PubMed]

42. Di Puccio, F.; Mattei, L. Biotribology of artificial hip joints. *World J. Orthop.* **2015**, *6*, 77. [CrossRef] [PubMed]

43. Tudor, A.; Laurian, T.; Popescu, V.M. The effect of clearance and wear on the contact pressure of metal on polyethylene hip prostheses. *Tribol. Int.* **2013**, *63*, 158–168. [CrossRef]

 *materials*

Article

The Mechanical Properties and In Vitro Biocompatibility of PM-Fabricated Ti-28Nb-35.4Zr Alloy for Orthopedic Implant Applications

Wei Xu [1,†], Ming Li [1,†], Cuie Wen [2], Shaomin Lv [1], Chengcheng Liu [1], Xin Lu [1,3,4,*] and Xuanhui Qu [1,3,4,*]

1   Institute for Advanced Materials and Technology, University of Science and Technology Beijing, Beijing 100083, China; xuweicool@126.com (W.X.); g20159107@xs.ustb.edu.cn (M.L.); lsmleon@163.com (S.L.); liucc1988@163.com (C.L.)
2   School of Engineering, RMIT University, 3083 Melbourne, Australia; cuie.wen@rmit.edu.au
3   Beijing Key Laboratory for Advanced Powder Metallurgy and Particulate Materials, University of Science and Technology Beijing, Beijing 100083, China
4   Beijing Laboratory of Metallic Materials and Processing for Modern Transportation, University of Science and Technology Beijing, Beijing 100083, China
*   Correspondence: luxin@ustb.edu.cn (X.L.); quxh@ustb.edu.cn (X.Q.); Tel.: +86-10-8237-7286 (X.L.); +86-10-6233-2700 (X.Q.)
†   These authors contributed equally to this work.

Received: 12 March 2018; Accepted: 29 March 2018; Published: 30 March 2018

check for
updates

Abstract: A biocompatible Ti-28Nb-35.4Zr alloy used as bone implant was fabricated through the powder metallurgy process. The effects of mechanical milling and sintering temperatures on the microstructure and mechanical properties were investigated systematically, before in vitro biocompatibility of full dense Ti-28Nb-35.4Zr alloy was evaluated by cytotoxicity tests. The results show that the mechanical milling and sintering temperatures have significantly effects on the density and mechanical properties of the alloys. The relative density of the alloy fabricated by the atomized powders at 1500 °C is only 83 ± 1.8%, while the relative density of the alloy fabricated by the ball-milled powders can rapidly reach at 96.4 ± 1.3% at 1500 °C. When the temperature was increased to 1550 °C, the alloy fabricated by ball-milled powders achieve full density (relative density is 98.1 ± 1.2%). The PM-fabricated Ti-28Nb-35.4Zr alloy by ball-milled powders at 1550 °C can achieve a wide range of mechanical properties, with a compressive yield strength of 1058 ± 35.1 MPa, elastic modulus of 50.8 ± 3.9 GPa, and hardness of 65.8 ± 1.5 HRA. The in vitro cytotoxicity test suggests that the PM-fabricated Ti-28Nb-35.4Zr alloy by ball-milled powders at 1550 °C has no adverse effects on MC3T3-E1 cells with cytotoxicity ranking of 0 grade, which is nearly close to ELI Ti-6Al-4V or CP Ti. These properties and the net-shape manufacturability makes PM-fabricated Ti-28Nb-35.4Zr alloy a low-cost, highly-biocompatible, Ti-based biomedical alloy.

Keywords: Ti-28Nb-35.4 alloy; powder metallurgy; ball milling; mechanical properties; biocompatibility

1. Introduction

Metallic biomaterials, which are the materials of choice for orthopedic implants, are fundamental for improve the quality of life and longevity of human beings [1,2]. Many different metallic materials have been used in a variety of applications in the medical field. Specifically, they are used for internal support and biological tissue replacements such as joint replacement, dental roots, orthopedic fixation and stents [3]. The common metals and alloys that are being utilized for biomedical applications include: stainless steels, Co-based alloys, and Ti-based alloys. Ti-based alloys have dominated in the human hard tissue repair and dentistry fields due to high strength, low elastic modulus, and excellent in vivo corrosion and biocompatibility [4]. However, despite their attractive performance, there are still challenges facing their application in artificial joints [5]. These concerns are mainly related to bio-toxicity, the need for lower elastic modulus and better mechanical strength. For instance, Al elements in extra-low interstitial (ELI) Ti-6Al-4V (hereafter all compositions are given in wt %), Ti-5Al-2.5Fe and Ti-6Al-7Nb [6–8], which is the most widely applied Ti alloy, are considered to be related to some health-related problems, including Alzheimer disease and neuropathy [9,10]. In addition, the elastic modulus (~110 GPa) of these alloys is still considerably higher than those of the cortical bones (~30 GPa), which results in severe 'stress shielding' for implantation failures [11,12]. Hence, considerable amount of effort has been exeerted to develop Al- and V-free lower-modulus β-Ti alloys.

Notable examples include the United States Food and Drug Administration (FDA)-approved proprietary alloys of Ti-13Nb-13Zr and Ti-12Mo-6Zr-2Fe and the non-proprietary alloy of Ti-15Mo. In particular, Ti-Nb-Zr alloys have received significant attention due to their excellent mechanical compatibility and biocompatibility [13–16]. The addition of Nb to Ti can stabilize the β phase and results in improved mechanical properties, which also results in improved wear resistance and corrosion resistance [17,18]. Furthermore, the addition of Zr helps in obtaining the solid solution required for achieving the hardness [19,20]. In addition, as reported by references [21,22], the elastic modulus in Ti-Nb and Ti-Zr alloys decreases with an increase in the content of the Nb and Zr within certain limits. Based on these results, Wen et al. [23] designed the new Ti-Nb-Zr alloys using the d-electron design method combined with the molybdenum equivalence (Mo_{eq}) and electron-to-atom ratio (e/a) approaches. The results show that the alloys have an excellent combination of mechanical properties and biocompatibility. However, at present these alloys are manufactured by the conventional route (i.e., ingot metallurgy in addition to wrought processing and machining with up to 90% being scrapped), which leads to a high manufacturing cost. Hence, in order to reduce costs and enhance the utilization rate of materials, the powder metallurgy (PM) technique is introduced. PM is an advanced net-shape technique which particularly suits to large volume production and can reduce processing steps, hence reducing cost [24–27]. It has already been used to synthesize Ti-based alloys by many researchers. For example, Sharma et al. [28] used a powder metallurgy route consisting of mechanical alloys (MA) of the TiH_2-Nb powder mixture and spark plasma sintering (SPS) to produce Ti-40 mass% Nb alloys. Jia et al. [29] successfully obtained the Ti-22Al-25Nb alloy by PM. Mendes et al. [30] produced the alloy Ti-27Nb-13Zr with low Young's modulus by PM using powders produced by the hydride-dehydride (HDH) process.

In this study, the Ti-28Nb-35.4Zr alloy is fabricated by PM in order to obtain Ti-based alloy with excellent mechanical compatibility and biocompatibility in addition to further reducing the manufacturing cost. The effects of sintering temperatures and mechanical milling on density, microstructure and mechanical properties are investigated systematically, before the in vitro biocompatibility of the full dense Ti-28Nb-35.4Zr is evaluated preliminarily. We aimed to establish a necessary understanding of the low-cost PM-fabricated Ti-28Nb-35.4Zr alloy for orthopedic implant applications.

2. Experimental and Methods

2.1. Materials and Sample Preparation

Atomized Ti-28Nb-35.4Zr powders (purity $\geq$ 99.9%, 75 $\leq$ particle size $\leq$ 150 μm) was supplied by the Wen group (RMIT University, Melbourne, Australia), who fabricated the powders using continuous inert gas atomization without the crucible method. The atomized Ti-28Nb-35.4Zr powders were subjected to ball milling for 30 min using a three-dimensional vibration ball milling machine (HSVM, Nanjing Chishun Science and Technology Co., Ltd., Nanjing, China). The frequency of the vibration ball milling machine was 1400 r/min, while the ball-to-powder weight ratio was 3:1.

The materials of the balls and jars were stainless steel and GCr15 bearing steel, respectively. The milling process was carried out in a high-purity argon atmosphere, with 2 wt % stearate used as the process control agent (PCA). Atomized powder (AP) and ball-milled powder (BMP) were cold-pressed into cylindrical compacts under 450 MPa. Then isothermal sintering was carried out in the argon (Ar) protection environment and implemented in two steps. Specimens were initially heated to 1000 °C for 2 h at 5 °C/min, before being heated at 2 °C/min to six different temperatures between 1200 °C and 1550 °C for 2 h. This was followed by furnace cooling to room temperature to obtain samples.

2.2. Materials Characterization

The density was measured by the Archimedes method and the relative density was calculated by the following formula:

$$\text{Relative density} = \text{the density/the theoretical density} \tag{1}$$

where the theoretical density is 6.36 g/cm^3. The hardness was carried out using the HDI-1875 Rockwell hardness tester. Five points were measured, before the average value was calculated. X-ray diffraction (XRD) was performed using a Dmax-RB X-ray diffractometer (Cu Kα, λ = 0.15406 nm, Rigaku, Tokyo, Japan). A JSM-6510V (JEOL, Tokyo, Japan) scanning electron microscope (SEM) was used to analyze the powder morphology of the received atomized powders and ball-milled powders and sintered microstructure. Compression specimens with a gauge size (φ) of 3 mm × 5 mm were fabricated by electric discharge machining and the specimen surface was polished with SiC papers. The compression test was performed on an Instron machine (Instron, Boston, MA, USA) at the strain rate of 2 × 10^{-3} s^{-1} at room temperature. The compressive yield strength and elastic modulus were calculated from the engineering stress strain curves.

2.3. In Vitro Biocompatibility Testing

Cytotoxicity tests were carried out with murine osteoblast cells (MC3T3-E1) to examine the in vitro biocompatibility of the Ti-28Nb-35.4Zr alloy. For comparison, the cast ELI Ti-6Al-4V and CP-Ti were studied simultaneously. The cells were cultured in Dulbecco's modified Eagle's medium (DMEM, Shanghai solarbio Bioscience and Technology Co., Ltd., Shanghai, China) containing 10% fetal bovine serum (FBS), 100 U/mL penicillin, and 100 μg/mL streptomycin at 37 °C under a humidified atmosphere of air containing 5% CO$_2$. The metal samples were cut into discs of 10 mm in diameter and 1 mm in thickness via electric discharge machining, before the surface was polished with SiC papers (grit 400 down to 5000). After this, the samples were cleaned ultrasonically and sterilized for further use.

In the cytotoxicity test, the extracts were obtained based on the international standard ISO 10993-5 [31]. The cells were incubated by an extraction medium in 96-well plates at the density of 5000 cells per 100 μL. At the desired times (day 1, day 2 and day 3), cells were observed by the optical microscope (LEXT OLS4000, Olympus, Tokyo, Japan), while the 10 μL MTT solution (Shanghai solarbio Bioscience and Technology Co., Ltd., Shanghai, China) was added to each well and were incubated for 4 h. After this, 100 μL of dimethyl sulfoxide (DMSO, Shanghai solarbio Bioscience and Technology Co., Ltd., Shanghai, China) was added to each well, before being incubated for a further 5 min. The absorbance was recorded by a multimode detector on a Synergy HT (BioTek, Winooski, VT, USA) at a wavelength of 570 nm. The cell viability ratio (CVR) was calculated by the formula as follows:

$$CVR = (OD_{570nm} \text{ in experimental extract}/OD_{570nm} \text{ in control extract}) \times 100\% \tag{2}$$

Based on the international standard ISO 10993-5 [31], the cytotoxic level was divided into six groups: 0: ≥100%; I: 75–99%; II: 50–74%; III: 25–49%; IV: 1–24%; V: ≤1%. Cell viability ratio were analyzed using one-way analysis of variance (ANOVA, P < 0.05) followed by the Tukey honestly significant difference (HSD) post-hoc test. P < 0.05 was considered to be statistically significant.

3. Results and Discussion

3.1. Raw Powder Characterization

Figure 1 depicts the representative SEM micrographs of atomized and ball-milled powders. It can be seen that the atomized powders (Figure 1a) have a typical spherical shape. The particle size of the powders varies from 30.2 to 160.5 μm with an average particles size of approximately 80.5 μm. The morphology of the Ti-Nb-Zr particles after 30 min of ball milling is shown in Figure 1b. The particle size of the ball-milled powders varies from 5.1 to 30.5 μm with an average particle size of approximately 15.2 μm. The shape of the powders becomes irregular, which makes further powder treatment easier (green compaction) as irregular powder particles have higher compressibility and green strength [32].

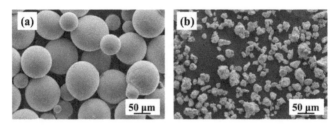

Figure 1. SEM images (800×) of the powders: (a) Atomized powders; and (b) Ball-milled powders.

Figure 2 shows the XRD patterns of atomized and ball-milled powders. It can be seen that atomized powders consist of a single β-Ti, while the ball-milled powders consist of β-Ti and TiO_2. TiO_2 mainly emerges because there is some oxygen introduced during the process of ball milling. In addition, compared with the atomized powders, the diffraction spectrums of the ball-milled powders shows an obvious broadening and moves to low angle, which is associated with the reduction in grain size, increase in lattice distortion and instrumental effects [33].

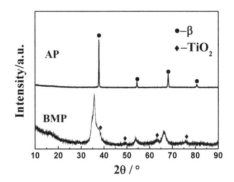

Figure 2. XRD patterns of atomized and ball-milled powders.

3.2. As-Sintered Density

Figure 3 shows the variation in relative density of samples with respect to sintering temperatures and powders. The relative density strongly depends on both sintering temperatures and powders. For the atomized powder, with an increase in the sintering temperature the relative density of the alloys increases gradually, but it is difficult to achieve a high relative density. The relative density of the alloy sintered at 1500 °C is only 83.1 ± 1.8%. Compared to the samples fabricated by atomized powder, the relative density of the samples fabricated by milled powder at 1500 °C rapidly increases to 96.4 ± 1.3%. When the sintering temperature increases to 1550 °C, the relative density of the alloy reaches 98.1 ± 1.2%.

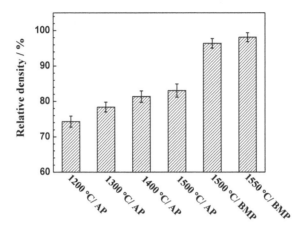

Figure 3. Relative density of Ti-28Nb-35.4Zr alloy prepared with different sintering temperatures and powders.

Figure 4 shows the optional images of Ti-28Nb-35.4Zr prepared with different sintering temperatures and powders. As shown in Figure 4a–d, the porosity of the samples fabricated by atomized powders decreases gradually with an increase in the sintering temperature. However, there are many pores on the surface of the samples even when they are sintered at 1500 °C. As shown by Figure 4e,f, the porosity of alloy prepared by milled powder reduces significantly, while the alloy becomes close to full density when sintered at 1550 °C. This is mainly associated with the sintering process of powder particles. After milling, the average particle sizes of the powders become smaller and hence, the powders have a high interface energy, which makes the sintering neck form more easily. Therefore, compared with the samples fabricated by atomized powder, the alloys prepared by milled powders have a higher density and therefore higher relative density.

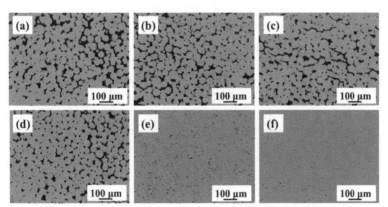

Figure 4. Optical images of Ti-28Nb-35.4Zr prepared with different sintering temperatures and raw powders: (**a**) 1200 °C/AP; (**b**) 1300 °C/AP; (**c**) 1400 °C/AP; (**d**) 1500 °C/AP; (**e**) 1500 °C/BMP and (**f**) 1550 °C/BMP.

3.3. As-Sintered Microstructure

Figure 5 shows the XRD patterns of Ti-28Nb-35.4Zr alloy fabricated by different powders at different sintering temperature. It can be seen that there is no significant difference of XRD patterns of Ti-28Nb-35.4Zr alloy fabricated by different powders and sintering temperatures. The clear diffraction peaks suggest that the samples all have similar phase compositions and mainly consist of single β phases.

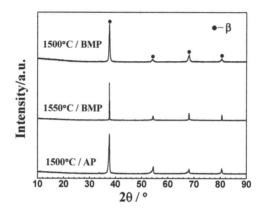

Figure 5. XRD patterns of Ti-28Nb-35.4Zr alloy prepared with different temperatures and powders.

Figure 6 displays the SEM micrographs of Ti-28Nb-35.4Zr alloy samples sintered at different temperatures with different powders, which was taken under 800× magnification. It can be seen that some pores exist in the alloy fabricated by the atomized powder at 1500 °C and the original powder particle boundaries can be observed clearly. However, the alloy prepared by milled powder at 1500 °C has a few pores, while the alloy becomes nearly fully dense with no obvious pores when the sintering temperature increases to 1550 °C. In addition, as shown in Figure 6b,c, the samples fabricated by the milled powder show a similar microstructure, which is consistent with the XRD results. Both samples consist of a single β phase and the average grain size is about 100 ± 10 μm.

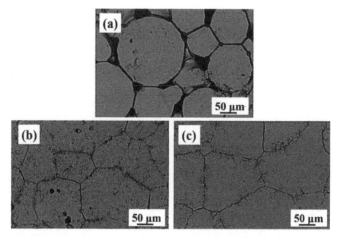

Figure 6. SEM images (800×) of Ti-28Nb-35.4Zr alloy prepared with different sintering temperatures and powders: (**a**) 1500 °C/AP; (**b**) 1500 °C/BMP and (**c**) 1550 °C/BMP.

3.4. Room-Temperature Mechanical Properties

Figure 7 shows compressive engineering stress-strain curves of the Ti-28Nb-35.4Zr alloy fabricated by different powders at different sintering temperatures. As shown in Figure 7, the alloys exhibit similar stress–strain behavior and no fracture was observed during their compression process, demonstrating that the alloys exhibit considerable elastic–plastic deformation ability. Compression tests were stopped after reaching a strain value of about 50%.

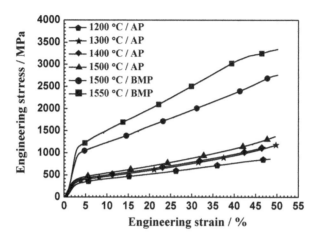

Figure 7. Compressive stress–strain curves of Ti-28Nb-35.4Zr alloy prepared with different sintering temperatures and powders.

The compressive yield strength, elastic modulus and hardness of Ti-28Nb-35.4Zr alloy fabricated by different powders at different sintering temperatures are shown in Figure 8. The hardness, compressive yield strength and elastic modulus of the Ti-28Nb-35.4Zr alloy fabricated atomized powder generally increase with an increase in the sintering temperature, which were in the range of 27.3–39.2 HRA, 273–332 MPa and 15.6–24.0 GPa, respectively. Compared with the alloy fabricated by the atomized powder, the mechanical properties of the alloy prepared by milled powder are significantly improved. The hardness of the alloy sintered at 1550 °C reaches 65.8 ± 1.5 HRA (Figure 8a) while the elastic modulus and compressive yield strength is 50.8 ± 3.9 GPa and 1058 ± 35.1 MPa (Figure 8b), respectively. This is superior to the mechanical properties fabricated by the cold-crucible levitation melting (compressive yield strength of 730 MPa [23]). The improvement in the compressive yield strength of the BMP sintered samples stems from two reasons. The first reason is that there is some oxygen introduced during the process of ball milling, which can also can be seen from the powders XRD results for the powders. In addition, compared with the samples fabricated by cold-crucible levitation melting, the PM-fabricated samples have smaller grains, uniform microstructure and less defects, which can also improve the strength.

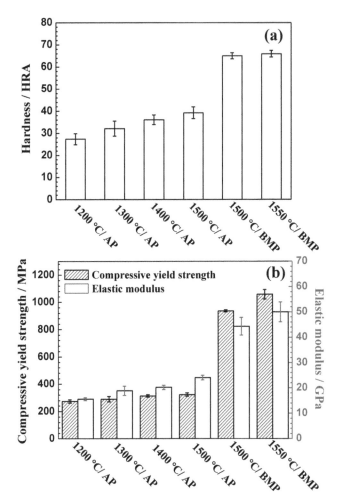

Figure 8. Mechanical properties of Ti-28Nb-34.5Zr alloy prepared with different sintering temperatures and powders: (**a**) Hardness; (**b**) Compressive elastic modulus and yield strength.

3.5. In Vitro Biocompatibility

In vitro biocompatibility of the Ti-28Nb-35.4Zr alloy sintered at 1550 °C by the milled powder was evaluated by cell tests. The CVR was calculated by Equation (2). The control group has a CVR of 100%, which is accepted as a reference for determining the cell viability radio of samples [34]. Figure 9 shows the results of CVR using MC3T3-E1 cells grown in Ti-28Nb-35.4Zr alloy, ELI Ti-6Al-4V and CP-Ti extracts for 1, 2 and 3 days. It can be seen that the average CVR value of all the experimental alloys was above 99%. The Ti-28Nb-35.4Zr alloy exhibits a slightly higher average CVR value than ELI Ti-6Al-4V while was similar to CP Ti. However there were no statistically significant differences among them (P > 0.05). According to ISO 10993-5 [31], the cytotoxicity grade of Ti-28Nb-35.4Zr alloy and CP Ti is 0–I grade while that of ELI Ti-6Al-4V is 0–I grade. Consequently, the PM-fabricated Ti-28Nb-35.4Zr alloy shows good in vitro biocompatibility to MC3T3-E1 cells.

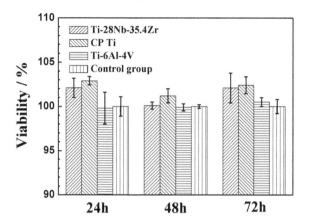

Figure 9. MTT result of MC3T3-E1 cells cultured with extracts of Ti-28Nb-35.4Zr, CP Ti and ELI Ti-6Al-4V.

Figure 10 displays the number and morphology of MC3T3-E1 cells cultured with extracts of Ti-28Nb-35.4Zr ELI Ti-6Al-4V and CP-Ti for different periods (1 day, 2 days and 3 days). It can be seen that the morphologies of all cells are similar for different periods. After a 1-day culture, cells are basically adhered to the well and the shape is spindle. With a prolonged culture time, cells cultured by different Ti alloys become denser and stretched, although there are no significantly changes in MTT.

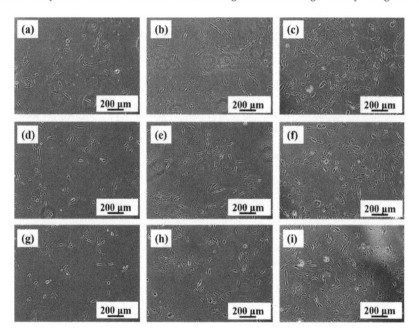

Figure 10. The number and morphology of MC3T3-E1 cells cultured with extracts of Ti-28Nb-35.4Zr ((**a**): 24 h; (**b**): 48 h; (**c**): 72 h), CP Ti ((**d**): 24 h; (**e**): 48 h; (**f**): 72 h) and ELI Ti-6Al-4V ((**g**): 24 h; (**h**): 48 h; (**i**): 72 h).

The in vitro biocompatibility of an implant lies not only in the toxicity of its alloys elements, but also in its corrosion resistance. Ti-Nb-Zr alloys are formed by the solid solutions of pure Ti, Nb and Zr, and maintain a high corrosion resistance and low toxicity [35]. Further, the corrosion resistance of Ti-Nb-Zr alloys is superior to that of CP Ti and ELI Ti-6Al-4V [36]. Spontaneously formed oxides, made up of TiO_2, Nb_2O_3 and Zr_2O_3, will provide a bio-inert layer in aggressive body fluids [37]. The enrichment of TiO_2, Nb_2O_3 and Zr_2O_3 on the surface suppressed the dissolution of Ti, Zr and Nb as ions. Hence, the Ti-28Nb-35.4Zr alloys show high cell viability during the MTT assays. Coupled with the low elastic modulus, high compressive yield strength and net-shape manufacturability, it thus suggests that PM-fabricated Ti-28Nb-35.4Zr alloy may hold promise for the development of next class of low-cost, highly-biocompatible, Ti-based biomedical alloys.

4. Conclusions

(1) Ti-28Nb-35.4Zr alloy with a high relative density and uniform microstructure can be obtained from milled powder via PM. After milling, the average particle sizes of powders was 15.2 μm, and when sintered at 1550 °C, the relative density reaches 98.1 ± 1.2%. The alloy fabricated by milled powders at 1550 °C is characterized by the single β phase.

(2) The alloys fabricated by the ball-milled powder at 1550 °C can achieve a high mechanical properties with the compressive yield strength of 1058 ± 35.1 MPa, the elastic modulus of 50.8 ± 3.9 GP, and the hardness of 65.8 ± 1.5 HRA. This is superior to the mechanical properties fabricated by the cold-crucible levitation melting technique.

(3) From in vitro cytotoxicity tests, the Ti-28Nb-35.4Zr alloy fabricated by milled powder at 1550 °C has no adverse effects on cell proliferation, and the cytotoxicity level is ranked as 0 grade, which is similar to the existing biomedical materials of CP Ti or ELI Ti-6Al-4V.

(4) Based on its demonstrated net-shape manufacturability by PM, high compressive yield strength, low elastic modulus and excellent in vitro biocompatibility, the PM-fabricated Ti-28Nb-35.4Zr can be considered as an attractive orthopedic implant alloy.

Acknowledgments: This research work is supported by Fundamental Research Funds for the Central Universities (FRF-GF-17-B39).

Author Contributions: Wei Xu, Ming Li and Xin Lu conceived and designed the experiments; Wei Xu and Ming Li performed the experiments; Wei Xu and Ming Li analyzed the data and wrote the paper; Chengcheng Liu, Shaomin Lv, Cuie Wen, Xin Lu and Xuanhui Qu revised the manuscript; all authors discussed and approved the final manuscript.

Conflicts of Interest: The authors declare no conflict of interest.

References

1. Okulov, I.V.; Volegov, A.S.; Attar, H.; Bönisch, M.; Ehtemam-Haghighi, S.; Calin, M.; Eckert, J. Composition optimization of low modulus and high-strength TiNb-based alloys for biomedical applications. *J. Mech. Behav. Biomed.* **2016**, *65*, 866–871. [CrossRef] [PubMed]

2. Attar, H.; Bönisch, M.; Calin, M.; Zhang, L.C.; Scudino, S.; Eckert, J. Selective laser melting of in situ titanium-titanium boride composites: Processing, microstructure and mechanical properties. *Acta Mater.* **2014**, *76*, 13–22. [CrossRef]

3. Dai, N.W.; Zhang, J.X.; Chen, C.; Zhang, L.C. Heat treatment degrading the corrosion resistance of selective laser melted Ti-6Al-4V alloy. *J. Electrochem. Soc.* **2017**, *164*, C428–C434. [CrossRef]

4. Biesiekierski, A.; Lin, J.; Li, Y.; Ping, D.H.; Yamabe-Mitarai, Y.; Wen, C.E. Investigations into Ti-(Nb,Ta)-Fe alloys for biomedical applications. *Acta Biomater.* **2016**, *32*, 336–347. [CrossRef] [PubMed]

5. Ehtemam-Haghighi, S.; Prashanth, K.G.; Attar, H.; Chaubey, A.K.; Cao, G.H.; Zhang, L.C. Evaluation of mechanical and wear properties of Ti-xNb-7Fe alloys designed for biomedical applications. *Mater. Des.* **2016**, *111*, 592–599. [CrossRef]

6. Lu, S.L.; Qian, M.; Tang, H.P.; Yan, M.; Wang, J.; StJohn, D.H. Massive transformation in Ti-6Al-4V additively manufactured by selective electron beam melting. *Acta Mater.* **2016**, *104*, 303–311. [CrossRef]

7. Siqueira, R.P.; Sandim, H.R.Z.; Hayama, A.O.F.; Henriques, V.A.R. Microstructural evolution during sintering of the blended elemental Ti-5Al-2.5Fe alloy. *J. Alloys Compd.* **2009**, *476*, 130–137. [CrossRef]

8. Polyakova, V.V.; Semenova, I.P.; Polyakov, A.V.; Magomedova, D.K.; Huang, Y.; Langdon, T.G. Influence of grain boundary misorientations on the mechanical behavior of a near-αTi-6Al-7Nb alloy processed by ECAP. *Mater. Lett.* **2017**, *190*, 256–259. [CrossRef]

9. Flaten, T.P. Aluminium as a risk factor in Alzheimer's disease, with emphasis on drinking water. *Brain Res. Bull.* **2001**, *55*, 187–196. [CrossRef]

10. Lima, P.D.; Vasconcellos, M.C.; Montenegro, R.C.; Bahia, M.O.; Costa, E.T.; Antunes, L.M.; Burbano, R.R. Genotoxic effects of aluminum, iron and manganese in human cells and experimental systems: A review of the literature. *Hum. Exp. Toxicol.* **2011**, *30*, 1435–1444. [CrossRef] [PubMed]

11. Wang, X.J.; Xu, S.Q.; Zhou, S.W.; Xu, W.; Leary, M.; Choong, P.; Qian, M.; Brandt, M.; Xie, Y.M. Topological design and additive manufacturing of porous metals for bone scaffolds and orthopaedic implants: A review. *Biomaterials* **2016**, *83*, 127–141. [CrossRef] [PubMed]

12. Xu, W.; Lu, X.; Zhang, B.; Liu, C.C.; Lv, S.M.; Yang, S.D.; Qu, X.H. Effects of Porosity on Mechanical Properties and Corrosion Resistances of PM-Fabricated Porous Ti-10Mo Alloy. *Metals* **2018**, *8*, 188. [CrossRef]

13. Lee, T.; Park, K.T.; Dong, J.L.; Jeong, J.; Sang, H.O.; Kim, H.S.; Park, C.H.; Lee, C.S. Microstructural evolution and strain-hardening behavior of multi-pass caliber-rolled Ti-13Nb-13Zr. *Mater. Sci. Eng. A* **2015**, *648*, 359–366. [CrossRef]

14. Niinomi, M. Fatigue performance and cyto-toxicity of low rigidity titanium alloy, Ti-29Nb-13Ta-4.6Zr. *Biomaterials* **2003**, *24*, 2673–2683. [CrossRef]

15. Zhang, L.C.; Klemm, D.; Eckert, J.; Hao, Y.L.; Sercombe, T.B. Manufacture by selective laser melting and mechanical behavior of a biomedical Ti-24Nb-4Zr-8Sn alloy. *Scr. Mater.* **2011**, *65*, 21–24. [CrossRef]

16. Niinomi, M.; Nakai, M.; Hieda, J. Development of new metallic alloys for biomedical applications. *Acta Biomater.* **2012**, *8*, 3888–3903. [CrossRef] [PubMed]

17. Zhou, F.Y.; Wang, B.L.; Qiu, K.J.; Lin, W.J.; Li, L.; Wang, Y.B.; Nie, F.L.; Zheng, Y.F. Microstructure, corrosion behavior and cytotoxicity of Zr-Nb alloys for biomedical application. *Mater. Sci. Eng. C* **2012**, *32*, 851–857. [CrossRef]

18. Li, S.J.; Yang, R.; Li, S.; Hao, Y.L.; Cui, Y.Y.; Niinomi, M.; Guo, Z.X. Wear characteristics of Ti-Nb-Ta-Zr and Ti-6Al-4V alloys for biomedical applications. *Wear* **2004**, *257*, 869–876. [CrossRef]

19. Kobayashi, E.; Ando, M.; Tsutsumi, Y.; Doi, H.; Yoneyama, T.; Kobayashi, M.; Hanawa, T. Inhibition effect of zirconium coating on calcium phosphate precipitation of titanium to avoid assimilation with bone. *Mater. Trans.* **2007**, *48*, 301–306. [CrossRef]

20. Hanawa, T.; Okuno, O.; Hamanaka, H. Compositional change in surface of Ti-Zr alloys in artificial bioliquid. *J. Jpn. Inst. Met. Mater.* **1992**, *56*, 1168–1173. [CrossRef]

21. Lee, C.M.; Ju, C.P.; Lin, J.H.C. Structure-property relationship of cast Ti-Nb alloys. *J. Oral Rehabil.* **2002**, *29*, 314–322. [CrossRef] [PubMed]

22. Ho, W.F.; Chen, W.K.; Wu, S.C.; Hsu, H.C. Structure, mechanical properties, and grindability of dental Ti-Zr alloys. *J. Mater. Sci.* **2008**, *19*, 3179–3186. [CrossRef]

23. Ozan, S.; Lin, J.X.; Li, Y.C.; Ipek, R.; Wen, C.E. Development of Ti-Nb-Zr alloys with high elastic admissible strain for temporary orthopedic devices. *Acta Biomater.* **2015**, *20*, 176–187. [CrossRef] [PubMed]

24. Pattanayak, D.K.; Matsushita, T.; Doi, K.; Takadama, H.; Nakamura, T.; Kokubo, T. Effects of oxygen content of porous titanium metal on its apatite-forming ability and compressive strength. *Mater. Sci. Eng. C* **2009**, *29*, 1974–1978. [CrossRef]

25. Niu, W.J.; Bai, C.G.; Qiu, G.B.; Wang, Q. Processing and properties of porous titanium using space holder technique. *Mater. Sci. Eng. C* **2009**, *506*, 148–151. [CrossRef]

26. Liu, Y.; Chen, L.F.; Tang, H.P.; Liu, C.T.; Liu, B.; Huang, B.Y. Design of powder metallurgy titanium alloys and composites. *Mater. Sci. Eng. A* **2006**, *418*, 25–35. [CrossRef]

27. Kipouros, G.J.; Caley, W.F.; Bishop, D.P. On the advantages of using powder metallurgy in new light metal alloy design. *Metall. Mater. Trans. A* **2006**, *37*, 3429–3436. [CrossRef]

28. Sharma, B.; Vaipai, S.K.; Ameyama, K. Microstructure and properties of beta Ti-Nb alloy prepared by powder metallurgy route using titanium hydride powder. *J. Alloys Compd.* **2012**, *656*, 978–986. [CrossRef]

29. Jia, J.; Zhang, K.; Liu, L.; Wu, F.Y. Hot deformation behavior and processing map of a powder metallurgy Ti-22Al-25Nb alloy. *J. Alloys Compd.* **2014**, *600*, 215–221. [CrossRef]

30. Mendes, M.W.D.; Ágreda, C.G.; Bressiani, A.H.; Bressiani, J.C. A new titanium based alloy Ti-27Nb-13Zr produced by powder metallurgy with biomimetic coating for use as a biomaterial. *Mater. Sci. Eng. C* **2016**, *63*, 671–677. [CrossRef] [PubMed]

31. ISO 10993-5. *Biological Evaluation of Medical Devices—Part 5: Tests for Cytotoxicity: In Vitro Methods*; American National Standards Institute: Arlington, VA, USA, 1999.

32. Samal, C.P.; Parihar, J.S.; Chaira, D. The effect of milling and sintering techniques on mechanical properties of Cu-graphite metal matrix composite prepared by powder metallurgy route. *J. Alloys Compd.* **2013**, *569*, 95–110. [CrossRef]

33. Wen, M.; Wen, C.E.; Hodgson, P.; Li, Y.C. Fabrication of Ti-Nb-Ag alloy via powder metallurgy for biomedical applications. *Mater. Des.* **2014**, *56*, 629–634. [CrossRef]

34. Li, Y.; Wen, C.W.; Mushahary, D.; Sravanthi, R.; Harishankar, N.; Pande, G.; Hodgson, P. Mg-Zr-Sr alloys as biodegradable implant materials. *Acta Biomater.* **2012**, *8*, 3177–3188. [CrossRef] [PubMed]

35. Wang, B.L.; Li, L.; Zheng, Y.F. In vitro cytotoxicity and hemocompatibility studies of Ti-Nb, Ti-Nb-Zr and Ti-Nb-Hf biomedical shape memory alloys. *Biomed. Mater.* **2010**, *5*, 044102. [CrossRef] [PubMed]

36. Robin, A.; Carvalho, O.A.S.; Schneider, S.G.; Schneider, S. Corrosion behavior of Ti-xNb-13Zr alloys in Ringer's solution. *Mater. Corros.* **2015**, *59*, 929–933. [CrossRef]

37. Robin, A.; Carvalho, O.A.S. Influence of pH and Fluoride Species on the Corrosion Behavior of Ti-xNb-13Zr Alloys in Ringer's Solution. *Adv. Mater. Sci. Eng.* **2013**, *1*, 1–10. [CrossRef]

 *materials*

Article

Wear Behaviours and Oxidation Effects on Different UHMWPE Acetabular Cups Using a Hip Joint Simulator

Saverio Affatato [1,*], Alessandro Ruggiero [2], Sami Abdel Jaber [1], Massimiliano Merola [2] and Pierangiola Bracco [3]

[1]   Medical Technology Laboratory, IRCCS—Rizzoli Orthopaedic Institute, Via di Barbiano, 1/10, 40136 Bologna, Italy; jaber@tecno.ior.it

[2]   Department of Industrial Engineering, University of Salerno, 84084 Fisciano, Italy; ruggiero@unisa.it (A.R.); mmerola@unisa.it (M.M.)

[3]   Chemistry Department and Nanostructured Interfaces and Surfaces (NIS) Centre, University of Turin, Via Giuria 7, 10125 Turin, Italy; pierangiola.bracco@unito.it

*   Correspondence: affatato@tecno.ior.it; Tel.: +39-051-636-6864; Fax: +39-051-636-6863

Received: 22 February 2018; Accepted: 14 March 2018; Published: 16 March 2018

Abstract: Given the long-term problem of polyethylene wear, medical interest in the new improved cross-linked polyethylene (XLPE), with or without the adding of vitamin E, has risen. The main aim of this study is to gain further insights into the mutual effects of radiation cross-linking and addition of vitamin E on the wear performance of ultra-high-molecular-weight polyethylene (UHMWPE). We tested four different batches of polyethylene (namely, a standard one, a vitamin E-stabilized, and two cross-linked) in a hip joint simulator for five million cycles where bovine calf serum was used as lubricant. The acetabular cups were then analyzed using a confocal profilometer to characterize the surface topography. Moreover; the cups were analyzed by using Fourier Transformed Infrared Spectroscopy and Differential Scanning Calorimetry in order to assess the chemical characteristics of the pristine materials. Comparing the different cups' configuration, mass loss was found to be higher for standard polyethylene than for the other combinations. Mass loss negatively correlated to the cross-link density of the polyethylenes. None of the tested formulations showed evidence of oxidative degradation. We found no correlation between roughness parameters and wear. Furthermore, we found significantly differences in the wear behavior of all the acetabular cups. XLPEs exhibited lower weight loss, which has potential for reduced wear and decreased osteolysis. However, surface topography revealed smoother surfaces of the standard and vitamin E stabilized polyethylene than on the cross-linked samples. This observation suggests incipient crack generations on the rough and scratched surfaces of the cross-linked polyethylene liners.

Keywords: vitamin-E stabilized PE; cross-linked PE; standard PE; hip simulator; FTIR analysis

1. Introduction

Ultra-high-molecular-weight polyethylene (UHMWPE) is a particular type of polyethylene (PE) with an exceptionally high molecular mass. It is a unique polymer with outstanding properties, in terms of chemical inertness, lubricity, impact, and abrasion resistance. Despite coming from a family of polymers with an extremely simple chemical composition, consisting of only hydrogen and carbon, UHMWPE shows a complex hierarchy of organizational structures at the molecular and supermolecular length scales [1]. Besides the molecular mass, the microstructure of the polymer also plays an important role in determining its physical, chemical, and mechanical properties. UHMWPE, as most polyethylenes, is a semi-crystalline polymer, composed of at least two interpenetrating phases: a crystalline, ordered phase, and an amorphous disordered one, possibly intercalated by a partially ordered interphase. Such a precise combination of chemical structure, molecular mass, and microstructure is at the basis of the peculiar balance of high mechanical properties and wear resistance that has made UHMWPE the material of choice in arthroplasty [2].

Thanks to its unique properties, UHMWPE is the most used material in hip joint replacement, being the soft insert coupled with harder materials (typically ceramics and metals). Such coupling shows very good tribological performances in terms of friction and wear [3,4]. Unfortunately, oxidative degradation can decrease its mechanical properties, leading to debris production and eventual osteolysis and implant loosening [5–7]. It is believed that wear of UHMWPE is to take place via plastic deformation of the polymer, with molecular alignment in the direction of motion that results in the formation of fine, drawn-out fibrils oriented parallel to each other. As a result of this arrangement, the UHMWPE wear surface may strengthen along the direction of sliding, while it weakens in the transverse direction [2]. Many efforts were done to improve the mechanical and molecular characteristic of the UHMWPE, as by cross-linking its molecular chains or by doping it with vitamin E [8]. Radiation cross-linking was demonstrated to improve the wear resistance of UHMWPE. On the other hand, irradiated polyethylene has also shown an unacceptably low oxidation stability [2]. As a consequence, stabilization strategies were developed in order to minimize post-irradiation oxidative ageing. Basically, two different strategies were adopted: one involved a thermal treatment of the polyethylene (re-melting or annealing), while the other included the addition of an anti-oxidant stabilizer. Oral and co-workers [9], suggested that the re-melting of the polyethylene reduces crystallinity and fatigue properties. Therefore, vitamin E was introduced to solve the oxidation problem [10]. Some authors [11–13], suggest to incorporate vitamin E in UHMWPE through blending the vitamin in the UHMWPE powder and then cross-link the blend through irradiation. With this process, the presence of vitamin E should protect the radiation-cross-linked polymer from oxidation, thus avoiding re-melting.

Preclinical evaluation of new biomaterials is necessary, and it could be considered as an extension of the risk analysis [14,15]. The wear performance of these improved biomaterials is often evaluated using hip joint simulators. Hip wear simulation tests are used since 40 years ago, and they represent a powerful system to assess the improvement in wear resistance before clinical use [16,17]. Some authors [18–20] observed a reduction in wear rate by the addition of vitamin E to highly cross-linked UHMWPE compared to conventional UHMWPE. However, there is a trade-off in determining the oxidation stability and the reduction in wear rate between the radiation dose applied for cross-linking and the amount of vitamin E incorporated into the polymer. Affatato and co-workers [21], using a hip joint simulator, found that the cross-linked polyethylene (XLPE) blended with vitamin E wore more than XLPE and conventional UHMWPE.

Each potential innovation has been accompanied by a great deal of pre-clinical trials, performed by researchers all over the world, often with very different methods and sometimes with contradictory results. In this regard, to go more in depth in the wear and oxidation behavior of new formulations of UHMWPE, we asked whether the addition of vitamin E on conventional UHMWPE could improve its wear performance in comparison with highly cross-linked polyethylene.

2. Materials and Methods

2.1. Specimens Tested

Four different batches of UHMWPE acetabular cups (32 mm inner × 50 mm outer dimensions; 6 specimens for each batch) coupled with 32 mm cobalt–chromium–molybdenum (CoCrMo) femoral heads have been investigated using a hip joint simulator. Three components of each batch run onto the simulator following a standardized procedure [22], another three acetabular cups for each type of material used were stored (non-loaded) in bovine calf serum to compensate for weight changes due to fluid absorption. All polyethylenes tested in this study were machined from polymer bars

Chirulen GUR 1020 (Polymax, Adler, Milan, Italy). Cross-linked acetabular cups were firstly γ-ray irradiated with 50 and 75 kGy (±10%), then thermally treated at 150 °C (re-melted), in order to remove free radicals formed during irradiation (hereinafter called XL-50 and XL-75). After these treatments, the cups were machined to their final shape. Similarly, vitamin E-containing (0.1% mass), UHMWPE acetabular cups (hereinafter called VE) were machined from a vitamin E-blended UHMWPE bars (Polymax, Adler, Milan, Italy). The UHMWPE that not received any treatment were called standard PE (STD). All the cups were then subjected to ethylene oxide sterilization (ETO). All polyethylene acetabular cups were pre-soaked for four weeks in a bath of deionized water prior the wear tests.

2.2. Experimental Wear Details

Wear test was performed using a 12-station hip joint simulator (IORSynthe, Bologna, Italy) [9]. The test was carried out applying the kinematic inputs and outputs as recommended by ISO 14242-1:2012. The simulator utilizes hydraulic actuators to apply the cyclic vertical compressive loads (oscillating between 300 and 3000 N). The lubricant used was 25 vol % newborn calf serum balanced with distilled water, with 0.2% (mass) sodium azide in order to retard bacterial growth, and 20 mM EDTA (ethylenediaminetetraacetic acid) to minimize precipitation of calcium phosphate. The mass loss of the cups was determined every 0.5 million cycles (Mc) using a microbalance (Sartorius Cubis Mse 225 S-000-DU, Goettingën, Germany) with a resolution of 0.01 mg and an uncertainty of 0.01 mg. Before the weighting operation, the specimens were cleaned from dust and possible debris using a dedicated detergent (Clean 70, Elma GmbH, Düsseldorf, Germany) in an ultrasonic bath maintained at 40 °C for 10 min. After rinsing, the cups were put back in the ultrasonic bath with deionized water for an additional 15 min. The cups were then dried with nitrogen gas. During any interruption of the test (every 500×10^5 cycles), the cups were stored in a closed, dust-free container at 70% of relative humidity. The test was re-started with fresh serum solution.

The test lasted five Mc, as recommended by the international guidelines [23], under environmental temperature conditions. The wear trend was determined from the mass loss of each acetabular cup, corrected by acetabular soak control. The wear rates, calculated from the steady-state slopes of the mass loss versus number of cycles, were obtained using least squares linear regression. The mass loss data were analyzed using a nonparametric Kruskall–Wallis (K–W) test; statistical significance was set at $p < 0.05$.

2.3. Surface Topography Characterization

The topographic analyses were performed using a PLu Neox profilometer (Sensofar, Terrassa, Spain), capable to gain three-dimensional images of a surface, operating either as confocal microscope or as white light interferometer with a vertical resolution, declared by the manufacturer, of less than 0.1 nm. In this study, we selected the confocal lens of $20\times$ magnifications, while the acquisition lengths were adjusted to compensate the inner curvature of the surfaces. Such a non-contact instrument was selected to measure the roughness of the polyethylene cups as optical techniques do not risk damaging the surface under investigation. On the other hand, a traditional roughness profilometer can scratch the surface of a soft material, such as the UHMWPE. The acquisition process followed an established procedure [24], where the polyethylene liners were first cleaned from debris, as described for the wear measurement, and—right before each acquisition—cleaned with ethanol and allowed to dry under a controlled environment in ambient air. In Figure 1, a schematic representation of the acquisition apparatus is shown.

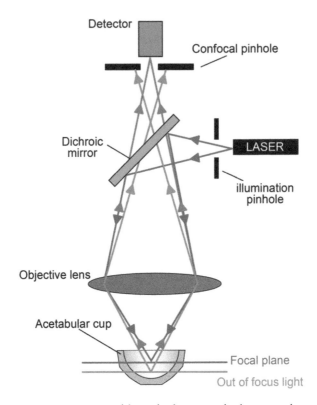

Figure 1. Schematic representation of the confocal apparatus for the topography acquisitions.

The topography characterization was realized on eight selected UHMWPE acetabular cups. In particular, we analyzed the most worn acetabular cups for each configuration, plus the corresponding check-control, to compare worn and unworn acetabular cups. The topographical acquisitions were realized to gain qualitative information on the surface condition of the samples after five Mc of test running. Furthermore, it was obtained a quantitative analysis in term of surface roughness. To do so, a Gaussian filter was applied according to the international guidelines in ISO 4287:1997 [25]. A cut-off wavelength of 80 μm was selected along with an evaluation length of 400 μm—equivalent to 5 times the cut-off. The roughness parameters selected were Ra, Rq, Rz, and Rt. Ra defines the absolute of the mean deviation of the irregularities from the mean line, over a sampling length. Rq is the standard deviation of the distribution of surface heights. Rz is the difference in height between the average of the five highest peaks and the five deepest valleys. Rt is the maximum height of the profile, defined as the vertical distance between the highest peak and the lowest valley along the measurement length.

2.4. FTIR Spectroscopy

All four different biomaterials were characterized by means of a Fourier Transformed Infrared Spectroscopy (FTIR) Microscope (Spectrum Spotlight 300, Perkin-Elmer, Shelton, CT, USA). A series of 180 μm thick slices was obtained from the cups cross section, using a PolyCuts Microtome (Leica Microsystem, Wetzlar, Germany) at 10 mm/s in air at room temperature. Line-scan spectra were collected on a 100×100 μm^2 area (resolution 4 cm^{-1}, 16 scans per spectrum), every 100 μm along the mapping direction, starting from the articulating surface towards the bulk. All spectra were normalized at 2020 cm^{-1} at an absorption of 0.05, corresponding to a film thickness of ca. 100 μm.

The combination band at 2020 cm^{-1}, associated with the twisting of CH$_2$, was used as an internal standard, since it can be regarded as unaffected by minor changes in the polymer structure. The molar concentration of trans-vinylene double bonds was calculated from the 965 cm^{-1} absorption bands, using the well-established molar absorptivity proposed by De Kock and Hol [26].

2.5. Determination of Cross-Link Density and Crystallinity

The cross-link density of each sample was quantified by gravimetric swelling. Small cylinders with diameter of 5 mm and approximate weight of 15 mg were cut from the control cups and immersed in 25 mL of xylene at 135 °C for 3 h to reach the equilibrium swelling. The initial weight and xylene uptake were used to calculate the swell ratio and the cross-link density, using a validated protocol [21].

The crystallinity of the test samples was determined using a differential scanning calorimetry (DSC 6-Perkin-Elmer, Waltham, MA, USA) at a heating rate of 10 °C/min. The sample weights varied around 5 mg. The heat of fusion was calculated by integrating the DSC endotherm from 60 to 160 °C. The crystallinity was calculated by normalizing the heat of fusion to the heat of fusion of 100% crystalline polyethylene (293 J/g) [27].

3. Results

All the polyethylene acetabular cups completed the planned five Mc. As showed in Figure 2, where the different polyethylene liners are compared, a higher mass loss rate for the standard polyethylene (STD) than for the other combinations was found. Close to this trend, the VE cups have a slightly smaller wear rate.

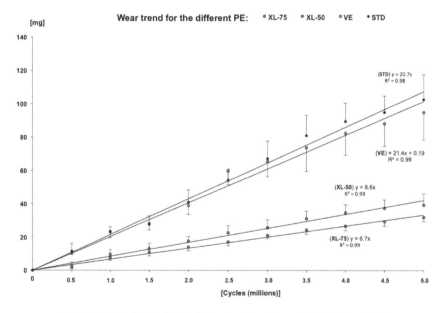

Figure 2. Wear behavior for the different configurations of polyethylene tested.

The polyethylene named XL-75 maintained the lowest mass loss than the other configurations during the whole test, as confirmed by the results of the post hoc test (Table 1). Statistical significant differences were observed between XL-75 vs UHMWPE and VE ($p = 0.015$ and $p = 0.02$, respectively). No statistically significant differences ($p > 0.05$) were observed between the other configurations of polyethylene cups using the K-S statistical test (Table 1).

Table 1. Cumulative mass loss (mean ± standard deviation) for the four sets of polyethylene (PE) acetabular cups tested. Values and statistical analysis performed using a Kruskall–Wallis nonparametric test.

Cycles (Mc)	Mean ± Standard Deviation (mg)				K–W Test	Post Hoc Test (p-Value)					
	STD	VE	XL-50	XL-75	(p-Value)	STD vs. VE	STD vs. XL-50	STD vs. XL-75	VE vs. XL-50	VE vs. XL-75	XL-50 vs. XL-75
0.5	11.4 ± 4.8	10.5 ± 1.2	2.7 ± 2.1	1.7 ± 0.7	0.037	0.821	0.079	0.036	0.047	0.020	0.734
1.0	22.9 ± 3.4	23.3 ± 3.8	9.6 ± 2.8	7.5 ± 1.6	0.035	0.910	0.089	0.024	0.070	0.017	0.571
1.5	27.6 ± 5.2	28.1 ± 3.9	13.0 ± 3.4	10.9 ± 2.3	0.038	0.910	0.070	0.031	0.054	0.024	0.734
2.0	40.9 ± 7.3	39.0 ± 5.5	17.4 ± 2.7	13.9 ± 2.4	0.030	0.910	0.089	0.013	0.113	0.017	0.428
2.5	54.1 ± 5.0	60.13 ± 8.8	22.5 ± 4.3	17.2 ± 2.4	0.026	0.571	0.174	0.031	0.054	0.007	0.428
3.0	67.0 ± 10.5	65.2 ± 9.1	25.7 ± 4.6	21.0 ± 1.9	0.035	0.910	0.070	0.017	0.089	0.024	0.571
3.5	80.8 ± 12.2	73.7 ± 14.2	31.2 ± 4.5	24.3 ± 2.6	0.030	0.910	0.089	0.013	0.113	0.017	0.428
4.0	89.4 ± 11.3	82.4 ± 13.3	34.6 ± 4.9	26.6 ± 2.8	0.030	0.910	0.089	0.013	0.113	0.017	0.428
4.5	94.8 ± 10.0	88.1 ± 13.6	37.3 ± 5.1	29.3 ± 2.7	0.030	0.910	0.089	0.013	0.113	0.017	0.428
5.0	102.6 ± 15.1	94.9 ± 16.7	39.3 ± 6.7	31.8 ± 2.6	0.032	0.910	0.079	0.015	0.100	0.020	0.496

Statistical significance: $p < 0.05$.

3.1. Surface Topography Analysis

Surface topography of the samples is shown in Figures 3 and 4, where contour images for each analyzed cup are presented. The different range of the height scales is influenced by the inner curvature of the cups. In Figure 3, the image shows the inner surfaces of the soak cups, which did not undergo wear simulation. The contours images of the topography present a similar pattern, typical of a clear UHMWPE surface for hip implants. In fact, the polymer has a lamellar shape, characterized by fine scratches, deriving from the polishing phase. Nevertheless, few long and transversal scratches are visible all over the images.

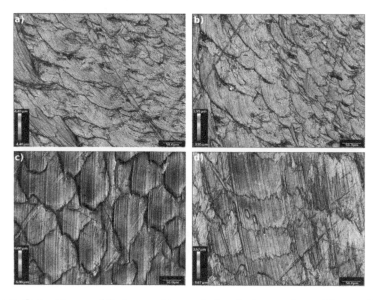

Figure 3. Contour images of the topographies acquired on the inner surfaces of four non-loaded specimens: (**a**) VE; (**b**) STD; (**c**) XL-75; (**d**) XL-50.

In Figure 4 are presented the contour topographies of the worn cups, after five million cycles of wear test. The images highlight that the inner surface of the cups is mainly characterized by the presence of grooves and scratches. In Figure 4a, a VE cup presents an almost smooth surface crossed by long and fine scratches; no signs of depth wear nor delamination are visible. In Figure 4b the inner surface of an STD cup is characterized by more frequent and larger grooves than on the VE. These paths are crossed in different directions. Also, signs of delamination are found along the border of these scratches. Figure 4c presents a worn area relative to a XL-75 cup. In this case, rough scratches are visible along with a deep groove (on the right side) and a worn crater (left side). In Figure 4d is shown a worn surface of a XL-50 cup, the surface presents many scratches in multiple directions and recurrent signs of wear along the side of these lines. Further worn images on the different configurations of the polyethylene tested, are shown in the supplementary section (Figure S1).

These qualitative analyses were combined with roughness measurements that are summarized in Figure 5. Ra values are equal to 0.09, 0.10, 0.25, and 0.45 μm respectively for the VE, STD, XL-75 and XL-50 cups. Rq is 0.19, 0.13, 1.5, and 2.1 μm, respectively, for VE, STD, XL-75, and XL-50; these values highlight how the deviation from the mean line is higher on the rougher surfaces of the cross-linked polyethylene. Rz and Rt provide evidence of the presence of elevate peaks and deep valleys along the measured profiles, and our analysis shows considerably higher values for the cross-linked polyethylene than the standard and the vitaminized ones. The values of Rz are 0.48, 0.57,

0.78, and 1.07 μm, respectively, for VE, STD, XL-75 and XL-50, whereas for the Rt parameters these are 1.69, 0.85, 22.7, and 30.7 μm, in the same order of correspondence.

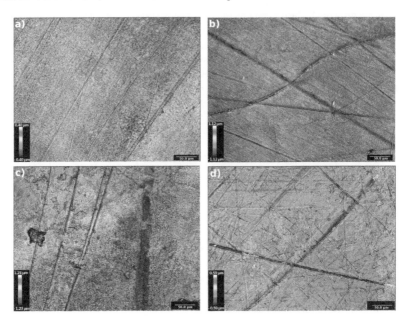

Figure 4. First set of contour images of the topographies acquired on the worn inner surfaces of four loaded specimens: (**a**) VE; (**b**) STD; (**c**) XL-75; (**d**) XL-50.

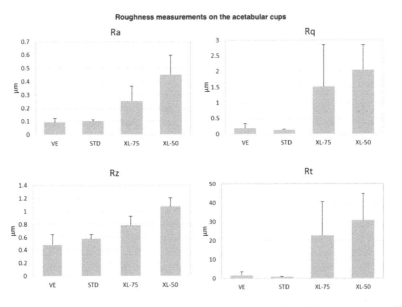

Figure 5. Roughness histograms of the worn surfaces on the mentioned cups. The parameters are Ra, Rq, Rz, and Rt.

3.2. FTIR Spectroscopy and Cross-Link Density Results

The physical–chemical characteristics of the control samples, resulting from FTIR, DSC, and cross-link density measurements, are summarized in Table 2.

Table 2. Physical–chemical characteristics of the control samples, resulting from FTIR, DSC, and cross-link density measurements.

Polyethylene	Crystallinity (%)	Trans-Vinylene (mmol/L)	Cross-Link Density (mol/dm^3)
STD	50.3 ± 0.9	-	-
VE	51.2 ± 1.2	-	-
XL-50	40.8 ± 1.0	5.2	0.132 ± 0.009
XL-75	35.5 ± 0.9	5.8	0.139 ± 0.010

Figure 6 shows the FTIR spectra of all materials. The VE sample spectrum does not show any significant difference from that of virgin UHMWPE (STD). On the contrary, the spectra of the irradiated samples show a decrease in the vinyl double bonds absorption at 909 cm^{-1} and the appearance of an additional absorption at 965 cm^{-1}, attributed to the formation of trans-vinylene double bonds, whose concentration is constant along the cup section, and increases with the radiation dose (see Table 2). No traces of oxidation products in the 1700 cm^{-1} area were observed in any of the samples (not shown). No significant differences in crystallinity were found between the STD and VE samples, while a significant decrease was observed in the crystallinity of the cross-linked polyethylenes. The cross-link density (ν_d) was measurable on the irradiated samples only (XL-50 and XL-75) and was found to increase as the radiation dose increases.

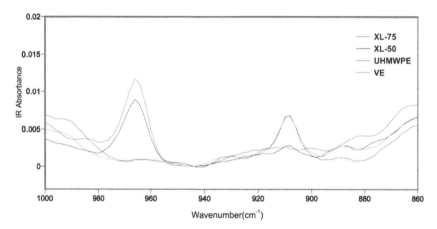

Figure 6. FTIR spectra of the investigated materials.

4. Discussion

The main purpose of this work was to characterize the wear performance of four different polyethylenes coupled with CoCrMo femoral heads using a 12-stations hip joint simulator for five million cycles. In particular, we asked whether the sole addition of vitamin E on conventional UHMWPE could improve its wear performance in comparison with high cross-linked polyethylene. We observed the highest wear rate on the STD polyethylene cups, followed closely by the VE, whereas the wear rate of XL-50 and XL-75 were sensibly lower than the former, but close to each other. In his research, McKellop et al. [6] found that the process of cross-linking highly improved the wear resistance, and the wear rate decreased markedly with increasing radiation dose, reaching a reduction of the 87% for the cup irradiated at 9.5 Mrad compared to the one irradiated at 3.3 Mrad. Affatato et al. [22] found that the wear of conventional UHMWPE was 40 times higher than for cross-linked polyethylene (XLPE), testing these materials against femoral heads of CoCrMo deliberately scratched.

The adding of vitamin E into the microstructure of conventional polyethylene used for hip components should prevent oxidative degradation and reduce the incidence of fatigue crack, confirming the results obtained by other authors [13]. The clinical consequence of the oxidation is an increased wear rate, starting approximately between 2 and 10 years postoperative [28].

Trans-Vinylene groups are known to be formed in UHMWPE upon irradiation, and can be used to assess the absorbed radiation dose [29]. The concentration of vinylene double bonds measured in our irradiated samples increases with the irradiation dose, as expected. At the same time, vinyl groups, normally present in virgin UHMWPE, are consumed as a consequence of irradiation, being involved in the formation of Y-shaped cross-links, as shown in Figure 7 [30].

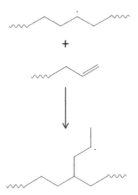

Figure 7. Mechanism of cross-linking of ultra-high-molecular-weight polyethylene (UHMWPE) through consumption of vinyl double bonds.

Accordingly, the FTIR analysis of both irradiated samples demonstrates consumption of the vinyl double bonds, while the cross-linking density measurements indicates that those samples have been cross-linked as a consequence of irradiation. Since cross-linking is known to increase the abrasion resistance, this explains the reduced wear, compared to the STD configuration, observed in the wear test. Nevertheless, both XLPE samples showed a significant decrease in crystallinity, and this is known to be correlated with a decrease in fatigue resistance, which creates concerns on their in vivo durability.

On the contrary, VE sample does not show any significant difference in the chemical and physical characteristics, compared to STD UHMWPE, indicating that the addition of vitamin E, which is intended as a stabilizer against oxidation, does not induce any modification to other properties. This also explains the similar behavior observed between VE and STD in the wear test.

From the topographical analysis, a very similar surface pattern of the unworn surfaces of the different polyethylene was found, verifying an almost equal initial condition for all of the cups that underwent the in vitro wear simulation. On the contrary, the inner surface of the worn samples showed large differences, both in terms of qualitative observation and in terms of rigorous measurement of the roughness. The contour images of the cross-linked polyethylene presented a more rough surface, with many signs of delamination and incipient cracks, which are symptoms of the fatigue wear phenomena [31]. On the other side, the non-cross-linked polyethylene presented a smoother surface than the XL ones, which suggests an abrasive wear occurrence that lead to a setting-in phase of the non-cross-linked polyethylene surfaces to the CoCrMo counterbody.

The roughness analysis emphasized these differences in the wear behavior of the polyethylene cups. In fact, higher roughness values on the cross-linked cups than on the standard and vitaminized ones were found. As grooves and scratches are assessed as origin points for surface-initiated micro-cracks, a high roughness value can be an indicator of the fatigue wear mechanism. As expressed in literature, cross-linking affects the compression and tension fatigue of the UHMWPE. Baker et al. [32] found growth of fatigue cracks under fully compressive cyclic loading in notched samples of UHMWPE. Cole et al. [33], in their experimental investigation on the fracture toughness and fatigue-crack resistance of two kinds of molded UHMWPE, found that "gamma irradiation decreased each material's resistance to fatigue-crack growth, and that the decreases were greater with increased irradiation dose".

5. Conclusions

All the acetabular cups studied in this work showed significant differences with respect to their wear behavior. Lower weight loss was exhibit by the XLPEs, which has potential for reduced wear and decreased osteolysis. The amount of weight loss after five million cycles in a hip joint wear simulator did not correlate with the roughness parameters, while cross-link density is strongly correlated with the wear resistance of UHMWPE. Nevertheless, the sensible higher presence of grooves and scratches on the cross-linked surfaces than on the standard ones is believed to be a plausible cause of fatigue wear. This, with the continuous cyclical compression of the element, could lead to an increased wear rate after elevate amount of working cycles.

Little is known or understood about the relationship between particle debris from PE wear and the biological response leading to osteolysis, and less is known about the role of cross-linking [12,28]. The promise of improved clinical performance in respect to osteolysis is a matter to be determined by long-term (>10 years) clinical follow-up studies. An increased understanding of the related biological processes and osteolysis is needed to provide the ability to better evaluate the STD, XLPEs, and VE performance in vivo.

Supplementary Materials: The following are available online at http://www.mdpi.com/1996-1944/11/3/433/s1, Figure S1: Supplementary figure with a second set of contour images of the topographies acquired on the worn inner surfaces of four loaded specimens: a VE; b STD; c XL-75; d XL-50. The observations are very similar to the ones described for Figure 4.

Acknowledgments: The authors would like to thank Barbara Bordini (Rizzoli Orthopaedic Institute) for her help with statistical analyses and Pasquale Cutino for his help during the topography acquisition. This work was supported by the Italian Program of Donation for Research "5 per mille", year 2014.

Author Contributions: S.A. and PB conceived and designed the experiments; S.A. and S.A.J. performed the wear tests; A.R. and M.M. performed the optical measurements and data analysis; S.A. and S.A.J. contributed materials and analysis tools; S.A., A.R., M.M. and P.B. wrote the paper.

Conflicts of Interest: The authors declare no conflict of interest.

References

1. Kurtz, S.M. A Primer on UHMWPE. *UHMWPE Biomater. Handb.* **2016**, 1–6. [CrossRef]
2. Bracco, P.; Bellare, A.; Bistolfi, A.; Affatato, S. Ultra-High Molecular Weight Polyethylene: Influence of the Chemical, Physical and Mechanical Properties on the Wear Behavior. A Review. *Materials* **2017**, *10*, 791. [CrossRef] [PubMed]
3. Ruggiero, A.; D'Amato, R.; Gómez, E.; Merola, M. Experimental comparison on tribological pairs UHMWPE/TIAL6V4 alloy, UHMWPE/AISI316L austenitic stainless and UHMWPE/AL2O3 ceramic, under dry and lubricated conditions. *Tribol. Int.* **2016**, *96*, 349–360. [CrossRef]
4. Ruggiero, A.; D'Amato, R.; Gómez, E. Experimental analysis of tribological behavior of UHMWPE against AISI420C and against TiAl6V4 alloy under dry and lubricated conditions. *Tribol. Int.* **2015**, *92*, 154–161. [CrossRef]
5. Brach del Prever, E.M.; Bistolfi, A.; Bracco, P.; Costa, L. UHMWPE for arthroplasty: Past or future? *J. Orthop. Traumatol.* **2009**, *10*, 1–8. [CrossRef] [PubMed]

6. McKellop, H.; Shen, F.W.; Lu, B.; Campbell, P.; Salovey, R. Development of an extremely wear-resistant ultra high molecular weight polyethylene for total hip replacements. *J. Orthop. Res.* **1999**, *17*, 157–167. [CrossRef] [PubMed]

7. Harris, W.H.; Muratoglu, O.K. A Review of Current Cross-linked Polyethylenes Used in Total Joint Arthroplasty. *Clin. Orthop. Relat. Res.* **2005**, 46–52. [CrossRef]

8. Affatato, S.; Freccero, N.; Taddei, P. The biomaterials challenge: A comparison of polyethylene wear using a hip joint simulator. *J. Mech. Behav. Biomed. Mater.* **2016**, *53*, 40–48. [CrossRef] [PubMed]

9. Oral, E.; Malhi, A.S.; Muratoglu, O.K. Mechanisms of decrease in fatigue crack propagation resistance in irradiated and melted UHMWPE. *Biomaterials* **2006**, *27*, 917–925. [CrossRef] [PubMed]

10. Oral, E.; Godleski Beckos, C.; Malhi, A.S.; Muratoglu, O.K. The effects of high dose irradiation on the cross-linking of vitamin E-blended ultrahigh molecular weight polyethylene. *Biomaterials* **2008**, *29*, 3557–3560. [CrossRef] [PubMed]

11. Azzi, A.; Stocker, A. Vitamin E: Non-antioxidant roles. *Prog. Lipid Res.* **2000**, *39*, 231–255. [CrossRef]

12. Oral, E.; Muratoglu, O.K. Vitamin E diffused, highly crosslinked UHMWPE: A review. *Int. Orthop.* **2011**, *35*, 215–223. [CrossRef] [PubMed]

13. Tomita, N.; Kitakura, T.; Onmori, N.; Ikada, Y.; Aoyama, E. Prevention of Fatigue Cracks in Ultrahigh Molecular Weight Polyethylene Joint Components by the Addition of Vitamin E. *J. Biomed. Mater. Res.* **1999**, *48*, 474–478. [CrossRef]

14. Trommer, R.M.; Maru, M.M. Review article Importance of preclinical evaluation of wear in hip implant designs using simulator machines. *Rev. Bras. Ortop.* **2016**, *52*, 251–259. [CrossRef] [PubMed]

15. Viceconti, M.; Affatato, S.; Baleani, M.; Bordini, B.; Cristofolini, L.; Taddei, F. Pre-clinical validation of joint prostheses: A systematic approach. *J. Mech. Behav. Biomed. Mater.* **2009**, *2*, 120–127. [CrossRef] [PubMed]

16. Affatato, S.; Spinelli, M.; Zavalloni, M.; Mazzega-Fabbro, C.; Viceconti, M. Tribology and total hip joint replacement: Current concepts in mechanical simulation. *Med. Eng. Phys.* **2008**, *30*, 1305–1317. [CrossRef] [PubMed]

17. Taddei, P.; Ruggiero, A.; Pavoni, E.; Affatato, S. Transfer of metallic debris after in vitro ceramic-on-metal simulation: Wear and degradation in Biolox® Delta composite femoral heads. *Compos. Part B Eng.* **2016**. [CrossRef]

18. Oral, E.; Christensen, S.D.; Malhi, A.S.; Wannomae, K.K.; Muratoglu, O.K. Wear resistance and mechanical properties of highly cross-linked, ultrahigh-molecular weight polyethylene doped with vitamin E. *J. Arthroplast.* **2006**, *21*, 580–591. [CrossRef] [PubMed]

19. Wannomae, K.K.; Christensen, S.D.; Micheli, B.R.; Rowell, S.L.; Schroeder, D.W.; Muratoglu, O.K. Delamination and adhesive wear behavior of alpha-tocopherol-stabilized irradiated ultrahigh-molecular-weight polyethylene. *J. Arthroplast.* **2010**, *25*, 635–643. [CrossRef] [PubMed]

20. Micheli, B.R.; Wannomae, K.K.; Lozynsky, A.J.; Christensen, S.D.; Muratoglu, O.K. Knee Simulator Wear of Vitamin E Stabilized Irradiated Ultrahigh Molecular Weight Polyethylene. *J. Arthroplast.* **2012**, *27*, 95–104. [CrossRef] [PubMed]

21. Affatato, S.; Bracco, P.; Costa, L.; Villa, T.; Quaglini, V.; Toni, A. In vitro wear performance of standard, crosslinked, and vitamin-E-blended UHMWPE. *J. Biomed. Mater. Res. A* **2012**, *100*, 554–560. [CrossRef] [PubMed]

22. Affatato, S.; Bersaglia, G.; Rocchi, M.; Taddei, P.; Fagnano, C.; Toni, A. Wear behaviour of cross-linked polyethylene assessed in vitro under severe conditions. *Biomaterials* **2005**, *26*, 3259–3267. [CrossRef] [PubMed]

23. *Implants for Surgery—Wear of Total Hip-Joint Prostheses—Part 1: Loading and Displacement Parameters for Wear-Testing Machines and Corresponding Environmental Conditions for Test*, 2nd ed.; ISO 14242-1:2012; International Organization for Standardization: Geneva, Switzerland, 2012.

24. Ruggiero, A.; Merola, M.; Affatato, S. On the biotribology of total knee replacement: A new roughness measurements protocol on in vivo condyles considering the dynamic loading from musculoskeletal multibody model. *Meas. J. Int. Meas. Confed.* **2017**, *112*. [CrossRef]

25. ISO 4287:1997. In *Geometrical Product Specifications (GPS)—Surface Texture: Profile Method—Terms, Definitions and Surface Texture Parameters*; International Organization for Standardization: Geneva, Switzerland, 1997.

26. De Kock, R.J.; Hol, P.A.H.M.; Bos, H. Infrared determination of unsaturated bonds in polyethylene. *Fresenius' Zeitschrift Für Analytische Chemie* **1964**, *205*, 371–381. [CrossRef]

27. Wunderlich, B.; Dole, M. Specific heat of synthetic high polymers. VIII. Low pressure polyethylene. *J. Polym. Sci.* **1957**, *24*, 201–213. [CrossRef]

28. Kurtz, S.M.; Dumbleton, J.H.; Siskey, R.S.; Wang, A.; Manley, M. Trace concentrations of vitamin E protect radiation crosslinked UHMWPE from oxidative degradation. *J. Biomed. Mater. Res. A* **2009**, *90*, 549–563. [CrossRef] [PubMed]

29. *Standard Test Method for Evaluating Trans-Vinylene Yield in Irradiated Ultra-High-Molecular-Weight Polyethylene Fabricated Forms Intended for Surgical Implants by Infrared Spectroscopy*; ASTM F2381-04; ASTM International: West Conshohocken, PA, USA, 2004.

30. Bracco, P.; Brunella, V.; Luda, M.P.P.; Zanetti, M.; Costa, L. Radiation-induced crosslinking of UHMWPE in the presence of co-agents: Chemical and mechanical characterisation. *Polymer* **2005**, *46*, 10648–10657. [CrossRef]

31. Nélias, D.; Dumont, M.L.; Champiot, F.; Vincent, A.; Girodin, D.; Fougéres, R. Role of Inclusions, Surface Roughness and Operating Conditions on Rolling Contact Fatigue. *J. Tribol.* **1999**, *121*, 240. [CrossRef]

32. Baker, D.A.; Hastings, R.S.; Pruitt, L. Study of fatigue resistance of chemical and radiation crosslinked medical grade ultrahigh molecular weight polyethylene. *J. Biomed. Mater. Res.* **1999**, *46*, 573–581. [CrossRef]

33. Cole, J.C.; Lemons, J.E.; Eberhardt, A.W. Gamma irradiation alters fatigue-crack behavior and fracture toughness in 1900H and GUR 1050 UHMWPE. *J. Biomed. Mater. Res.* **2002**, *63*, 559–566. [CrossRef] [PubMed]

Review

3D Printing of Bioceramics for Bone Tissue Engineering

Muhammad Jamshaid Zafar, Dongbin Zhu * and Zhengyan Zhang *

School of Mechanical Engineering, Hebei University of Technology, Tianjin 300130, China;
jamshaid.zafer@yahoo.com
* Correspondence: zhudongbin@hebut.edu.cn (D.Z.); zzy@hebut.edu.cn (Z.Z.)

Received: 4 September 2019; Accepted: 8 October 2019; Published: 15 October 2019

Abstract: Bioceramics have frequent use in functional restoration of hard tissues to improve human well-being. Additive manufacturing (AM) also known as 3D printing is an innovative material processing technique extensively applied to produce bioceramic parts or scaffolds in a layered perspicacious manner. Moreover, the applications of additive manufacturing in bioceramics have the capability to reliably fabricate the commercialized scaffolds tailored for practical clinical applications, and the potential to survive in the new era of effective hard tissue fabrication. The similarity of the materials with human bone histomorphometry makes them conducive to use in hard tissue engineering scheme. The key objective of this manuscript is to explore the applications of bioceramics-based AM in bone tissue engineering. Furthermore, the article comprehensively and categorically summarizes some novel bioceramics based AM techniques for the restoration of bones. At prior stages of this article, different ceramics processing AM techniques have been categorized, subsequently, processing of frequently used materials for bone implants and complexities associated with these materials have been elaborated. At the end, some novel applications of bioceramics in orthopedic implants and some future directions are also highlighted to explore it further. This review article will help the new researchers to understand the basic mechanism and current challenges in neophyte techniques and the applications of bioceramics in the orthopedic prosthesis.

Keywords: bioceramics; additive manufacturing; scaffolds; bone tissue engineering

---

1. Introduction

Additive manufacturing or 3D printing has got attention in scaffold design and manufacturing for tissue engineering applications. Initially, this technique was developed by Sachs et al., to create the ink-jet freestyle printing towards the latter part of the 20th century [1]. Later on, it was extended in tailoring the perfect scaffolds on its user-friendly capabilities, which considered the transformation of computer aided design (CAD) information to a rapid and reliable production line of constructs with the coveted material, porosity, and measurements [2,3]. Moreover, it showed a time and cost-efficient potential coupled with interconnected structures, specifically hard tissue deformity regeneration.

Recently, clinical preliminaries and contextual analyses revealed its resounding accomplishments in the field of orthopedic bioengineering. While this procedure has shown significant potential, specific difficulties tend to enhance patient-particular scaffolds for standard acknowledgment in regenerative medicine [4–6].

During the past few decades, many advanced biomaterials were introduced in the biomedical field including different ceramic materials for the skeletal repair and reconstruction. These materials in the field of medical implants are often referred to as "bioceramics" [7]. Bioceramics are peculiar in nature due to their exceptional biological and osteoinduction properties. These materials are specific for scaffolds due to capability to create propagation, self-adhesion, distinction and bone tissues regeneration [8]. Furthermore, excellent chemical and mechanical properties such as better osteoconductivity, superior wear resistant and biocompatibility enabled them as a substitute for bone restoration, [9,10]. It can anticipate that bioceramics have a future due to increasing bone replacement operation per year due to increasing aging population [11].

The clinical importance of AM ceramic scaffold design and implantation envelops an invaluable method for quick and reliable production of hard tissue substitution replica of the biological context of natural bone [12]. In view of the way that customized scaffold can be prepared that suits an individual patient's skeletal imperfection, layer-by-layer sintering is regarded as a lucrative discipline for the utilization of ceramic-based bone substitutes in regenerative medicine [13]. Besides, utilizing AM ceramic scaffolds as medication conveyance systems, it is becoming more attractive and relevant to the bioengineering environment [14–16].

This article is divided into six sections; Section 1 details the bioceramics potential, Section 2 offers an overview of the AM techniques used to fabricate ceramic parts. Section 3 presents achievements in the production of hydroxyapatite (HA); Section 4 depicts about tricalcium phosphate (TCP) and Section 5 describes about bioactive glass (BG) using different AM techniques. Section 6 concludes some important findings with some current challenges and future opportunities in this field.

2. Additive Manufacturing Technologies to Produce Ceramic Parts

Additive manufacturing has been classified into two major classes such as acellular and cellular techniques for biomaterials. The cellular category includes the printing of live cells, while the acellular category does not consider any type of live cells in printing. Figure 1 shows different acellular AM techniques for biomaterials that have been classified as per recommendations of American Society for Testing of Materials (ASTM) [17]. The major AM techniques employed in the processing of bioceramics have been discussed in the following section.

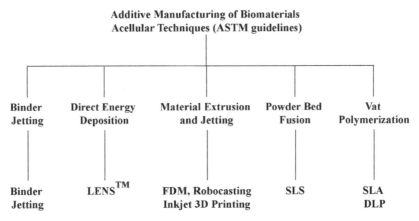

Figure 1. Different types of acellular techniques for biomaterials additive manufacturing (AM).

2.1. Binder Jetting

Binder jetting technique was developed in the early 1990s at Massachusetts Institute of Technology (MIT) [18]. Figure 2 depicts the schematic of binder jetting. In this technique, the binder is selectively used from powder bed to create 3D objects. Binder jetting is a valuable technique for printing powder materials [19,20]. The particle size of the powder has a key influence on powder flowability in binder jetting.

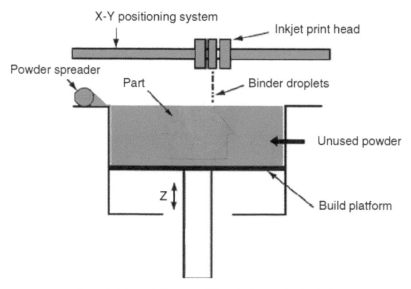

Figure 2. Schematic diagram of binder jetting mechanism [18].

For dry binder jetting, large size particles are preferred due to its outstanding flowability and less surface area. The powder size not only affects the flowability but also significantly affects the quality of the final product. Numerous researchers have reported less surface roughness using fine powder in the binder jetting. The effect of the powder shape is less, as compared to the powder size. However, spherical shape powders have better flowability and lesser friction compared to faceted powders [21–23].

2.2. Direct Energy Deposition (DED)

Direct energy deposition-based AM techniques uses energy into a small region to simultaneously deposit, melt and solidify the material such as wire or powder [24]. The direct energy source can be electrical, or laser beam can be used to melt the metal, ceramics or composite materials. Laser assisted Direct Deposition techniques such as laser cladding, laser engineered net shaping (LENS™), and laser melt injection are common examples of this technique.

In ceramic Direct Energy Deposition (DED), the printing head of the apparatus contains a nozzle that feeds ceramic powder particles to the focal point of the laser beam. The ceramic powder melts and solidifies in layer-wise fashion on a substrate [25]. Figure 3 is the schematic illustration of LENS [26].

The major advantages of DED are better compatibility with a wide range of biomaterial viscosities, higher resolution and greater cell density that provide higher control of cell-to-cell adhesions [27]. Besides these advantages, DED has many challenges such as, low speed, cost, high complexity and limited capability to manufacture heterogeneous tissue parts [28]. Due to these challenges, the usage of DED is very limited as compared to other AM techniques particularly in bone tissue engineering. The DED technology needs more research to enhance its productivity.

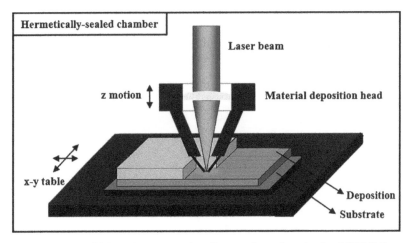

Figure 3. Typical Schematic representation of laser engineered net shaping (LENS) [26].

2.3. Material Extrusion and Jetting

Extrusion assisted additive manufacturing deposits a continuous layer by layer deposition of ceramic loaded paste to create 3D objects. Various terms are used to refer to this technology for instance, Fused Deposition of Ceramics (FDC), Robocasting, Extrusion Freeform Fabrication (EFF), Direct Ink Writing (DIW), Slurry Deposition, and Dispense Plotting [29].

In Fused Deposition, dense ceramic particles (up to 60 vol%) are spread into a wax or thermoplastic filament after which the flexible filament is partly melted and extruded from a moving deposition head onto a fixed worktable layer-by-layer. However, in robocasting, ceramic slurry is ejected from a precise nozzle to form a filament that is directly deposited in a designed pattern to create complex 3D objects in a layer-by-layer fashion [30].

In another research work, an indirect Fused Deposition Modeling (FDM) method was applied to prepare ceramic parts. At the preliminary stage, FDM was used to prepare a honeycomb shaped polymer structural mold. Secondly, the ceramic slurry was permeated into the polymer mold-sintering to remove the mold. The porous ceramics made a correct pore size and porosity through this technique [31].

Another technique named Extrusion-based bioprinting has also a greater potential in perspective of deposition and printing speed compared to other AM techniques. This technique is also beneficial to achieve better scalability in a shorter time [32], wide range flexibility of bioinks selection [33]. This is because developing new bioinks is a critical procedure for quick, sustainable and safe delivery of cells in a biomimetic microenvironment [32]. Besides many advantages, some complexities are also associated with this technology such as low resolution and shear stress effect on cells. The schematic of the process with part microstructure was shown in Figure 4.

The material jetting techniques are the "AM processes in which droplets of build material are selectively deposited" [17], that can be used to manufacture different kinds of ceramic parts. Inkjet 3D printing technology was among the first material jetting AM techniques that were employed for additive manufacturing of ceramic parts. It was developed by Sachs et al. in 1992 at MIT and defined as a process for the manufacturing of ceramic casting cores and shells using inkjet 3D printing [1]. Figure 5 shows the Schematic of ink-jet 3D printing [34].

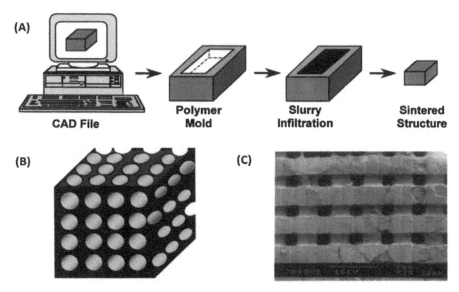

Figure 4. (**A**) Schematic of indirect Fused Deposition Modeling (FDM) processing of ceramic parts (**B**) Straight channels (**C**) Top view of sintered porous ceramic part [31].

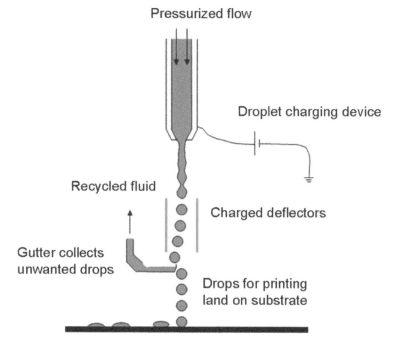

Figure 5. Schematic diagram depicts the basic working principle of ink-jet 3D printing [34].

2.4. Powder Bed Fusion

Powder bed fusion (PBF) technologies are among pioneer commercially used AM techniques created by the University of Texas USA. Selective laser sintering (SLS) based PBF technique [18], which melts the ceramic powder by laser energy source. The laser sintered the powder nearly to the melting point of the material to make each layer according to the given 3D design. The laser beam scans each new single layer of free-packed powder particles and consolidates them by sintering this process and proceeds in a layer-wise fashion to complete the final 3D object [35–37]. SLS is a powder bed fusion process have numerous applications in the bioengineering field such as to prepare customized products, biomedical implants as well as orthopedic implants [38]. The major disadvantage of SLS is the usage of higher temperatures that limits the insertion of biomaterial and cells into SLS scaffolds during the manufacturing process [39]. A schematic diagram illustrates the underlying operating system [40] of the powder bed fusion provided in Figure 6. While SLS technology is amended, the PBF method to increase machine efficiency.

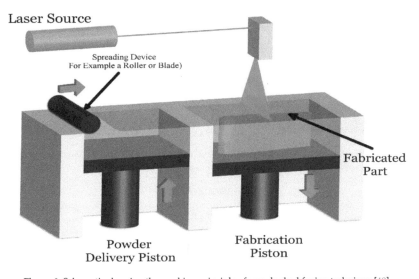

Figure 6. Schematic showing the working principle of powder bed fusion technique [40].

2.5. Vat Polymerization (SLA)

Vat Polymerization also known as stereolithography (SLA) is a promising AM technique to fabricate tissue scaffolds in the field of regenerative medicine. The SLA technique has exceptional control over porosity of scaffolds, pore sizes, design flexibility, and interconnectivity [41]. Despite excellent advantages, numerous researchers have highlighted several challenges in scaffold manufacturing such as, difficult in creating micron-sized scaffold due to over curing and layer thickness. In addition, some of the frequently used biomaterials in bone tissue engineering have shown compatibility with SLA due to limitations in viscosity, refractive index and stability [42].

Some other problems such as some SLA processes light pixels restrict in-plane microstructure construction. Although indirect SLA have overcome this problem, it is a costly, time and material consuming process [43]. Li et al. have used indirect stereolithography to manufacture microporous β-TCP. The resin molds were prepared through this technique and filled with filled with aqueous thermosetting ceramic suspension for ceramic gel casting. The heat treatment process was used to remove the molds. Results have concluded that TCP scaffolds after sintering have shown desired porosity, shape and higher strength were obtained [44].

Some other researchers have mentioned preparation of 3D objects by photo-curing a liquid resin through ultraviolet (UV) laser in a layerwise fashion [45,46]. The major advantage of this process includes better surface finish and accuracy [47]. A schematic of three different light sources used in stereolithography provided in Figure 7 [18]. Table 1 summarizing some basic bioprinting techniques for bone tissue engineering.

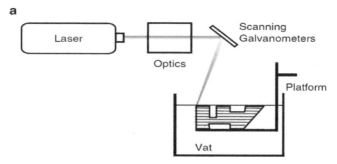

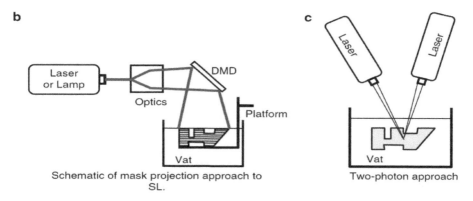

Figure 7. Schematic diagrams of three different techniques of photopolymerization [18].

Table 1. Summary of major 3D-bioprinting techniques for bone tissue engineering.

Technique	Principle	Advantages	Drawbacks
Inkjet	A liquid binding material is selectively deposited in layer-wise fashion into the powder bed to create three dimensional objects.	Ability to print biomaterials with low viscosity, high resolution, fast manufacturing speed, low cast	Intrinsic inability to deliver a continuous flow, low cell densities, lack of functionality for vertical objects
Extrusion	This process involves extruding the material in viscous form to create 3D objects	Capability to print variety of biomaterials, Capable of printing high cell densities	Applicable to viscous liquids only
Laser-assisted	In this technique, a laser beam stimulates a specified area of target to fabricate 3D objects	High resolution, capable of printing both solid and liquid phase biomaterials	High cost, low speed, high complexity, thermal damage due laser irritation
Stereolithography (SLA)	In this method an ultraviolet (UV) laser beam selectively hardens the photo-polymer resin to construct 3D models in layer-wise fashion	Nozzle free method, high cell viability, high accuracy, Printing time independent of complexity, high cell viability, high accuracy	UV light can cause toxicity to cells, during photo curing damage to cells, Applicable to photopolymers only

3. Additive Manufacturing of Bioceramics

In the last few decades, bioceramics have frequently been used in the restoration and replacements of injured tissues due to numerous advantages such as precise chemical composition, which has a vital role in the integration of hard and soft tissues [48,49]. Hydroxyapatite (HA) $Ca_{10}(PO_4)_6(OH)_2$ is one bioceramic to have frequently been employed as a scaffold material for bone tissue engineering owing to its exceptional biocompatibility and resemblance to natural bone material [50–52]. It is often combined with a biopolymer or bioceramics to enhance binding interaction and mechanical properties of the material during the AM process [53,54].

Beta tri-calcium phosphate (β-TCP) is a suitable material for craniofacial defects owing to its excellent biodegradability, wear resistance and chemical bonding with the bone tissues under all load bearing conditions [55,56]. The critical challenge for β-TCP is to maintain the sintering temperature of 1100 °C. Above this temperature, beta tri-calcium phosphate (β-TCP) transforms to alpha tri-calcium phosphate (α-TCP) that is soluble and chemically unstable, as compared to β-TCP [57,58].

In addition, bioglasses are also extensively used in hard tissue implants due to their excellent bonding capability with hard and soft both tissues. Bioglasess are also extremely helpful in upregulating the osteogenesis, however, their application in load bearing bone defects are very limited due to their high brittleness, low fracture toughness and mechanical strength [59–61]. Properties of some frequently used ceramic material for bone tissue engineering illustrated in Table 2.

Table 2. A brief review of ceramic materials and its properties used in 3D printing of Scaffolds.

Materials	Precursors	Properties
Hydroxyapatite (HA)	Poly (acrylic acid), photo-curable resin, polycaprolactone, poly (lactic acid) etc.	Higher biocompatibility, differentiation and proliferation, better cell adhesion
Tricalcium Phosphate (TCP)	Hydroxypropyl methylcellulose, polyethylenimine, polymethacrylate, etc.	In physiological environment better biocompatibility and degradation, lower compressive strength
Bioactive glasses alkali-free bioactive glass, 45S5 BG,13-93 bioactive glass, 6P53B glass	Polycaprolactone, methylcellulose, poly (lactic acid)	Improved bioactivity in vitro and in vivo for the bone tissue growth

The key factor affecting the performance of bioceramics is Ca to P ratio that affects the dissolution property. Calcium phosphates with lower Ca to P ratio (β-TCP) have higher solubility and acidic nature as compared to calcium-phosphate having high Ca to P ratio (HA) [62]. Table 3 shows that lower the Ca/P ratio higher the CaP dissolution [63]. Different bioceramics have been discussed in the following section such as hydroxyapatite, beta tri-calcium phosphate (β-TCP) and bioactive glass (BG) using different AM techniques. Figure 8 [64] shows complete process of bone tissue engineering.

Table 3. Characteristics of main CaPs used as bone substitutes and cements [63].

Name	Formula	Ca/P Ratio	Water Solubility at 25 °C, g/L
Monocalcium Phosphate			
Monohydrate (MCPM)	$Ca(H_2PO_4)_2, H_2O$	0.50	~18
Anhydrous (MCPA)	$Ca(H_2PO_4)_2$		~17
Dicalcium phosphate			
Dihydrate (DCPD)	$CaHPO_4, H_2O$	1.00	~0.088
Anhydrous (DCPA)	$CaHP_4$		~0.048
Tricalcium Phosphate			
Alpha α-TCP	$(\alpha)\ Ca_3(PO_4)_2$	1.50	~0.0025
Beta β-TCP	$(\beta)\ Ca_3(PO_4)_2$		~0.0005
Hydroxyapatite (HA)	$Ca_5(PO_4)_3OH$	1.67	~0.0003

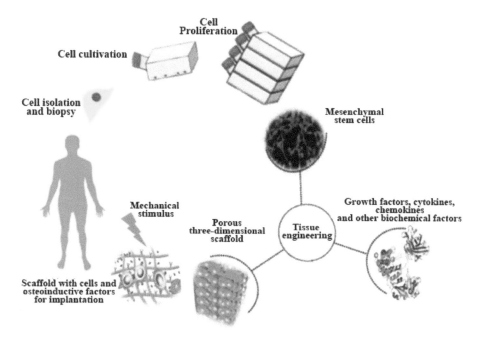

Figure 8. Schematic representation of bone tissue engineering [64].

3.1. Hydroxyapatite (HA)

Hydroxyapatite (HA) portrayed as $Ca_{10}(PO_4)_6(OH)_2$, encompasses almost 65% of the entire bone mass. It is less toxic and more stable, as compared to other calcium-phosphate due to their desirable Ca to P ratio of 1.67. The hydroxyapatite has major inorganic part of human bone and teeth to develop the properties and novel applications of bioceramics for hard tissue replacements [65–68]. Numerous researchers have reported HA scaffolds in the bone and teeth transplants [69–77].

Laser Stereolithography has been identified as one of the most effective and frequently used AM techniques to produce complex HA parts. Barry et al. have prepared HA-based oligocarbonate dimethacrylate (OCM-2) composite scaffolds using helium-cadmium (HeCd) based laser technology. The outcomes referred the laser-based HA scaffolds provided fortified cell attachment inside the scaffold. Through laser machining, toxic leftovers were removed effectively through supercritical carbon dioxide (scCO₂) to make scaffolds biocompatible. The HA based composite materials treated by scCO₂ showed better attachment of cells in both vivo and vitro studies [78]. In a very recent study, a bio-ink was prepared for 3D printing by dispersing two different types of hydroxyapatites, nano hydroxyapatite (nHA) and deproteinized bovine bone (DBB) into collagen. Thereby, a porous structure was created by 3D printing. The chemical and physical properties of the materials, including biocompatibility and effect on the osteogenic differentiation of the human bone marrow-derived mesenchyme stem cells (hBMSCs) were investigated. Both nHA/CoL and DBB/CoL Bio-inks were used to print biomimic 3D scaffolds effectively. The outcomes from this study showed that the two types of hydroxyapatite composites which help hBMSCs proliferation and differentiation proved to be a promising candidate for a 3D scaffold bio-ink [79].

Woeszn et al. fabricated microporous HA scaffolds having a pore size of 450 μm through stereolithography coupled with ceramic gelcating. A photosensitive liquid resin filled with water based thermosetting slurry was used in the mold. The mold resin and sintering were burnt to achieve the desired features. The final Scaffolds were seeded on MC3T3-E1 cells for 14 days under deep penetration of cells to achieve outstanding osteogenesis as shown in Figure 9 [80].

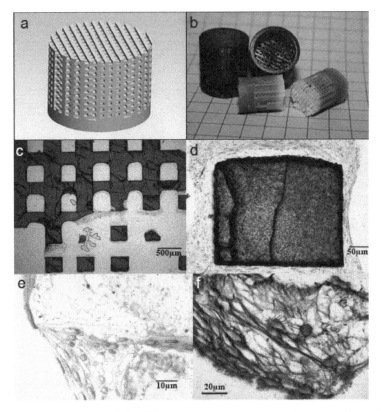

Figure 9. (**a**) Computer aided design (CAD) designed virtual structure of casting mold; (**b**) Resin casting molds manufactured by stereolithography and sintered hydroxyapatite (HA) structures; (**c**) SEM image of HA scaffold after culturing with MC3T3-E1 cells for 2 weeks, scaffold is visible in (dark grey) and cells (blue); (**d**) SEM image of strut of HA scaffold (grey), entirely attached with cells (blue/pink); (**e**) SEM image of microstructure of a crack between two struts, which was totally covered by MC3T3 cells (blue) and matrix created by the cells(pink); (**f**) SEM images of collagen produced by the cells (the microtome sectioning eliminates the mineral scaffold) [80].

The AM based extrusion process is also very common to manufacture HA scaffolds. The robocasting based extrusion process contains ceramic ink in the form of water-based viscous slurry deposited on a robotic nozzle in layer-wise fashion based on computer-aided design. The process contains high loading of HA particles to minimize the cracks and distortion during sintering. Saiz et al. have fabricated HA scaffolds with controlled pore sizes through robocasting extrusion to find the optimum sintering temperature. The hydroxyapatite slurry was prepared by mixing 40–50 vol.% of HA powder in distilled water, 1.5 wt% of Darvan C dispersant, (~7 mg/mL of solution) hydroxypropyl methylcellulose, an adequate amoqunt of polyethyleneimine (PEI) and at the end HNO_3 or HN_4OH to balance the pH level of the slurry. Results concluded that porous HA scaffolds manufactured with robocasting showed the sintering temperature should remain between 1100 °C~1200 °C and no phase change was observed for firing 1300 °C for 3 h. The characteristics of printed scaffolds through this technique have been presented in Figure 10 [81].

Keriquel et al. have successfully printed the nano-HA scaffold in the mouse calvaria defect model in vivo using laser-assisted additive manufacturing. The printed cells showed the existence of vivacious blood vessels after bone defect treatment. The outcomes of this study demonstrated that laser-assisted bioprinting have perfectly treated bone defects. Through literature numerous authors have mentioned potential of this technology could offer new perspectives to additive manufacturing for the practical applications of bone tissue engineering [82].

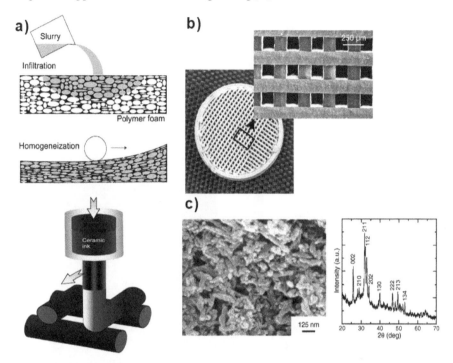

Figure 10. (a) Graphical representation of methods used in the manufacturing of porous ceramic scaffolds (b) Microstructure of HA scaffolds fabricated by robocasting (c) SEM micrograph and XRD of HA powders used in this process [81].

3.2. Tricalcium Phosphate (TCP)

Since the last two decades, beta tri-calcium phosphate (β-TCP) ceramic-based scaffolds have clinically accepted the bone graft replacement materials in several orthopedic and dental applications [83–89]. The TCP contains α, $\alpha\prime$ and $\beta\prime$ phases and Ca to P ratio is about 1.5. Cao et al. manufactured sphingosine 1-phosphate (S1P) coated β-TCP scaffold. Immunoregulation capability was tested on macrophages and rat bone marrow stromal cells of the coated scaffold was used to test osteogenic capability. The scaffold exhibit improved osteogenesis, better cell compatibility and also helpful to regulate the immune response as compared to traditionally manufactured scaffold. Figure 11 is the representation of 3D printed scaffold and its cell viability [90].

Bian et al. introduced a novel stereolithographic method to produce osteochondral beta-tricalcium phosphate/collagen scaffold. This bio-inspired scaffold manufactured by a combination of ceramic stereolithography (CSL) and gel casting using (β-TCP) and type-I collagen. Histological examination was performed to investigate the morphological properties between cartilage and bone. The obtained information from this examination were used to design biomimetic biphasic scaffolds. The pores size of β-TCP scaffolds varied between 700–900 μm with 50–65% porosity and compressive strength of 12 MPa. Physical locking formed by biomimetic transitional structure was used to achieve an adequate binding force among cartilage phase and a ceramic phase. The results concluded that CSL performed well in comparison with traditional techniques to get an ideal scaffold for bone tissue engineering applications [91].

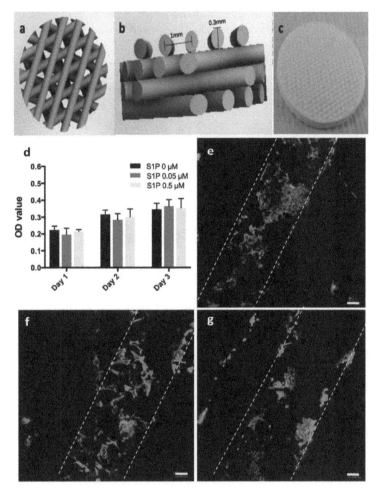

Figure 11. (**a–c**) Illustrates the schematic of 3D printed scaffold. (**d**) showing that there is no noticeable difference on viability of bone marrow-derived mesenchyme stem cells (BMSC) cells on additively manufactured scaffolds coated with S1P among the control group (S1P 0 mM) and other groups after 3 days. (**e**) S1P 0 mM group, (**f**) S1P 0.05 mM group, (**g**) S1P 0.5 mM group. Dyed blue area representing the cell nuclei and green area showing cytoskeletons. Edge of filaments showed by dotted lines [90].

In a recent study, Bose et al. have investigated the effect of Fe^{3+} and Si^{4+} dopants on the bio-mechanical properties of 3D printed β-TCP scaffold in a rat distal-femur for the period of 4, 8 and 12 weeks. The scaffold was fabricated by binder jetting technique using synthesized β-TCP powder. The outcomes from this analysis demonstrated that the addition of Fe^{3+} to TCP scaffold speed up the early stage bone restoration boosting type I collagen production. Si^{4+} doped TCP scaffold showed neovascularization after 12 weeks as shown in Figure 12. The finding from this study proved that ceramic powder-based scaffolds with improved chemistry has a promising future in bone defect restoration [92]. Tarafder et al. manufactured β-TCP scaffolds with 27%, 35% and 41% designed macroporosity with pore sizes of 500 μm, 75 μm and 1000 μm, respectively by 3D printing method. After that the scaffolds were sintered at the temperature of 1150 °C to 1200 °C in conventional and microwave furnaces to achieve mechanical strength. Microwave sintering heated scaffolds showed higher mechanical strength, as compared to conventional sintering.

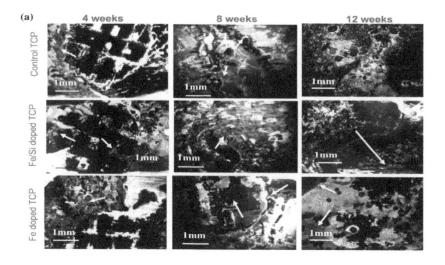

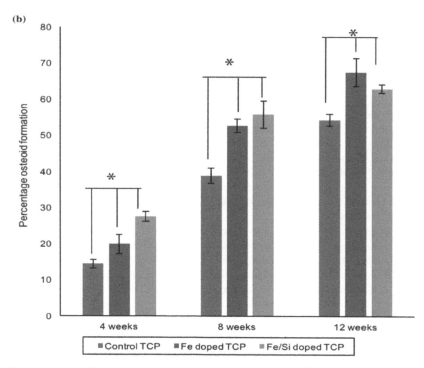

Figure 12. (a) Osteoid formation in pure and doped tricalcium phosphate (TCP) scaffolds after Modified Masson-Goldner trichrome staining after 4, 8, and 12 weeks. Black: prosthesis, orange and red: osteoid, bluish green: mineralized bone. Reddish-orange colors indicated by arrows showing new bone formation. Fe doping showed more bone mineralization as compared to others. (b) Histomorphic analysis showed Fe-Si doped TCP boosted initial stage osteoid formation for the period of 8 weeks and Fe doped TCP shows better mineralization of bone for 12 weeks of implantation [92].

The pore size was examined by Human osteoblast cells. This study showed that a decline in pore size from 1000 to 750 and 500 μm has increased the cell density. Histomorphology tests in femoral defects of Sprague-Dawley rats revealed that the existence of both micro and macro pores accelerated the new bone construction. It was concluded that additive manufactured TCP scaffolds have outstanding potential in hard tissue engineering applications [93].

3.3. Bioactive Glass (BG)

Bioactive glasses (BG) are the type of bioceramics that exist in both nonporous and solid forms. The bio glass contains silicon dioxide, sodium oxide, calcium oxide and phosphorous. Different types of bioglass have been created by varying the vol.% of these components [94]. Silicate part plays have an essential role in the biocompatibility of bioactive glasses. The bio glass with 45–52 vol.% silicate has ideal bone-graft bonding [95]. The 45S5-bioglass is a well-known commercially available extensively used in bone replacement [96]. However, bioglasses have also some limitations due to poor mechanical properties and brittleness that makes them unsuitable for load-bearing applications, internally brittle and deficient mechanical strength. However, several researchers have reported different additives, such as metal, polymer and ceramic to enhance the mechanical properties [97–101].

Recently, Nommeots-Nomn et al. robocasted bioglass scaffolds with a 150 μm interconnected pore size (41–43% porosity) and measured compressive strengths were 32–48 MPa. The network connectivity (NC) of these scaffolds is like the 45S5 bioglass. In this process, ICIE16 and PSrBG compositions were used comprising < 50 mol% SiO_2 to maintain the amorphous structure and to achieve the required NC closer to 45S5 bioglass. The manufactured scaffolds were compared with 13–93 vol.% composition bioglass. The comparison highlighted that 3D porous scaffolds have similar NC values with 45S5 bioglass using two low silica contents. In addition, Pluronic F-127 binder could be accepted as a universal binder for bioactive glasses regardless of their composition and reactivity. Results also showed that ICIE16 and PSrBG based scaffolds are highly reactive and significantly enhanced the bone regeneration speed [102].

Padilla et al. used calcined bioglass suspension to fabricate porous scaffolds through integrating the stereolithography and gel-casting method. A polymeric negative mold was used via stereolithography to cast bioglass suspension with Darvan-811 (sodium polyacrylate) as a dispersant. The slurry containing 50 vol% was heated at 1100 °C for 55 s and later it was polymerized. The negative mold was removed by heat treatment. The scaffolds containing interconnected 3D channels of 400–470 μm length and 1.4 μm of pore diameter. The results illustrated that the entire interconnected porous scaffold was created by this method [103].

Westhauser et al. inspected the osteo-inductive properties of different polymer coated 3D-45S5 bioglass scaffolds. These scaffolds are seeded with human mesenchymal stem cells (hMSC) implanted into immunodeficient mice. The gelatin, cross-linked gelatin, and poly (3-hydroxybutyrate-co-3-hydroxyvalerate) type coatings were used. histomorphometry and micro-computed topography analysis were performed to evaluate the new formation after eight weeks of implantation. Although, every bioglass scaffolds showed noticeable bone regeneration. However, gelatin-coated bioglass scaffolds showed highest cell formation in comparison with other coated-bioglasses, as shown in Figure 13 [104]. Some latest researches on 3D printing of bioceramics have been compiled in Table 4.

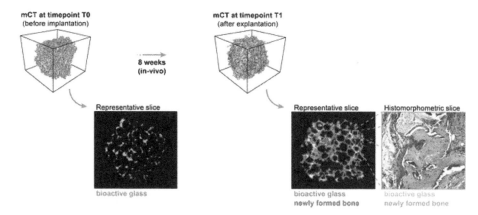

Figure 13. Histomorphometric and micro-CT analysis showing new bone development in polymer coated Bioglass (BG) scaffolds inserted in mice for eight weeks [104].

Table 4. Overview of 3D printed bioceramics for bone tissue engineering.

Material	Process	In Vivo/In Vitro Model	Key Findings	Ref.
HA + liquid sodium polyacrylate + photopolymer	A ball crusher was used to milled all the materials for 12 h to make a slurry with solid content of 10~60 wt%. The ceramic scaffold was fabricated by using digital light processing (DLP) technique	Mouse osteoblast precursor cells (MC3T3.E1) were cultured in the condition of α-MEM (10% fetal bovine serum 4% penicillin-streptomycin)	3D printed scaffold showed better biocompatibility, adhesion, differentiation and also able to promote osteoblast proliferation	[105]
Biphasic calcium phosphate (HA/β-TCP = 60:40) + HMPC+ polyethylenimine + ZrO$_2$	Extruded at pressure of 600 kPa with printing speed of 100 mm/min. Constructs were sintered at 1100 °C	Investigated on osteoblast like sarcoma cells for cytotoxicity and for differentiation potential of the scaffolds hMSCs cells were used	Better mechanical properties of scaffolds at 10% (*w/w*) of ZrO$_2$ was observed with improved BMP-2 expression.	[106]
β−TCP/polycaprolactone	β-TCP powder with 550 nm particle size were used to fabricate 350 μm pore size cylindrical scaffolds.	Composite scaffolds were tested using human fetal osteoblast cells (hFOB) for 3, 7 and 11 days of incubation period	Enhanced early bone formation and effective for controlled alendronate release	[107]
β−TCP/sphingosine 1-phosphate (SIP)	The scaffolds were printed in four layers and in different sizes to fit in 6-well and 12-well plates. Printed scaffolds were sintered at 1100 °C for 3 h.	Immunoregulation capability was investigated on macrophages and the osteogenic capability was tested on rat bone marrow stromal cells of the coated scaffolds.	Good biocompatibility, improved bone regeneration process	[90]
Bioactive glass/alginate	Composite scaffolds of type 13-93 bioactive glass (13-93 BG) and sodium alginate (SA) were prepared with mass ratio of 0:4, 1:4, 2:4 and 4:4 under mild conditions for bone regeneration.	The apatite mineralization abilities of the 13-93 BG/SA scaffolds were tested by soaking scaffolds in simulated body fluid (SBF), using 200 mL g^{-1} of scaffold mass, at 37 °C for 0 and 10 days.	Improved porosity and reduced shrinkage ratios	[108]
Bioglass (BG)/gelatin/cross linked-gelatin/ploy (3-hydroxybutyrate-co-3-hydroxyvalerate)	Three different types of 3D-polymer coated BG (45S5-type) scaffolds were fabricated by the well-established foam replica method and coated with the biopolymers.	Osteo-inductive properties of 3D-45S5 bioglass scaffolds were investigated by seeding human mesenchymal stem cells (hMSC) implanted into immunodeficient mice for the period of 8 weeks.	Under standard conditions biopolymer coated 3D 45S5 BG scaffolds have ability to induce bone formation. Gelation coated scaffolds showed the best results.	[104]

4. Application of Bioceramics in Orthopedic Implants

Natural bone has self-repair capability after the damage. The smaller fractures heal itself correctly, however segmental bone defects (SBDs) lead to permanent paralysis [109,110]. SBDs fractures treated with autologous bone graft technique requires harvesting of non-vital bone, such as, the iliac crest.

However, some complexities are also associated with bone grafting such as bone availability, the mismatch between harvested bone and affected site, morbidity of donor site results in poor integration [111]. Over the past three decades, a variety of synthetic materials have been introduced to overcome the complexities such as calcium phosphates (bioactive glasses) and hybrid bioceramics-polymer materials [112–115]. Table 5 showing different materials for bone tissue engineering.

Table 5. Additive manufacturing (AM) materials for bone prostheses.

Material	Binder	Layer Thickness	References
TCP	Aqueous based	20 μm	[93]
HA	-	100 μm	[116]
α/β-TCP modified with 5 wt% hydroxypropymethylcellulose	Water	100 μm	[117]
β-TCP, SiO$_2$-ZnO-dope β-TCP	Water based binder	20 μm (β-TCP) 30 μm (SiO$_2$-ZnO-doped β-TCP)	[118]
HA	α-n-butyl cyanoacrylate (NBCA)	-	[119]
TCP	20% (v/v) phosphoric acid	125 μm	[120]
TTCP/β-TCP	25% citric acid	100 μm	[121]
α-TCP	10 wt.% phosphoric acid	50 μm	[122]
HA/Maltrodextrin	Water based	175 μm	[123]
HA & Maltrodextrin/ apatite-wollastonite glass	Water based	100 μm	[124]

Roohani-Esfahani et al. fabricated glass-ceramic scaffolds with hexagonal pore structure via extrusion-based AM method shown in Figure 14. The fabricated scaffolds have 150 times greater strength compared to polymeric-composite scaffolds and five times greater than ceramic-glass scaffolds having same porosity. The study has shown that these scaffolds have excellent capability to load-bearing and segmental bone defects treatment [125].

Fierz et al. prepared HA based cylindrical scaffolds ranging from nanometer to millimeter with straight channels and micro-pores through n-HA granules, ink-jet 3D printing AM technique. The structure of 3D-printed scaffolds is almost similar to human cortical and cancellous bone. The histological analysis has confirmed that osteogenic-stimulated progenitor-based 3D-printed scaffolds are suitable for clinical use [126].

In another study, a robocasting technique was utilized to transport bone morphogenic protein 2 (BMP-2). HA slurry and polymethylmethacrylate (PMMA) microspheres were mixed together to achieve controlled microporosity. Resins were eliminated by sintering the scaffolds at 1300 °C for 2 h. Thereafter, 10 μg of bone morphogenic protein 2 was added to the microporous scaffolds in goat bone for in vivo characterization for 4 and 8 weeks. Outcomes from this study showed great potential for manufacturing HA scaffolds containing interconnected porosity. Furthermore, the existence of bone morphogenic protein 2 and micro porosity upgraded scaffold osteogenesis ability as illustrated in Figure 15 [127].

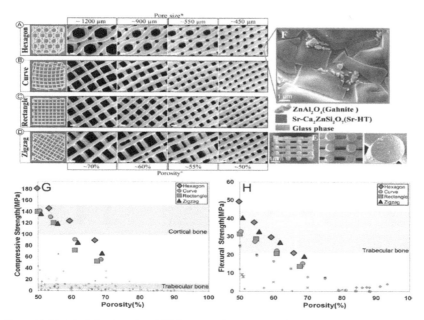

Figure 14. Models (left column) and SEM images of inspected scaffolds (scale bars: 500 μm unless stated otherwise); (**A**) Hexagonal; (**B**) curved; (**C**) rectangular and; (**D**) zigzag shape; (**E**) SEM images of fracture surface of a Sr-HT-Gahnite scaffold fabricated by robocasting; (**F,G**) the microstructure of Sr-HT-Gahnite scaffolds with distinct pore geometries vs porosity, and (**H**) flexural strength of Sr-HT-Gahnite scaffolds with hydroxyapatite and BG scaffolds [125].

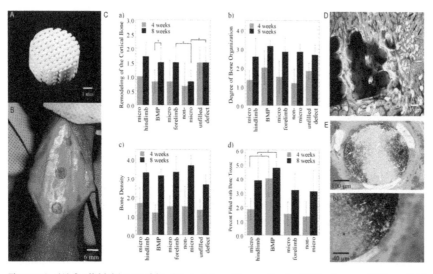

Figure 15. (**A**) Scaffold fabricated by directed deposition method. (**B**) The image of HA scaffold implantation in the metacarpal bone of goat. (**C**) BMP-2 and microporosity on cortical bone. (**D**) Image of BMP scaffold after 8 weeks representing the remodeling of the host bone, indicated by arrows. (**E-a**) and (**E-b**) are the images of histological section of micro hindlimb after 4 weeks indicating the staining of the microporous scaffolds at (**a**) low magnification and at (**b**) high magnification. Arrows indicate (1) stained and (2) unstained and (3) regions where staining extends into the scaffold [127].

Fielding et al. introduced (SiO$_2$/ZnO) doped three-dimensional composite TCP scaffolds with a pore size of 300 μm using binder jetting technique. The pure and (SiO$_2$/ZnO) doped 3D-printed TCP scaffolds implanted into a rat femur bone for the period of 6, 8 and 12 weeks to analyze the histomorphometry and Immunohistochemistry. Results have proved that combining SiO$_2$-ZnO dopants in TCP are best alternative to achieve osteoinductive properties of calcium phosphates (CaPs) for the clinical application of bone implants as shown in Figure 16 [118].

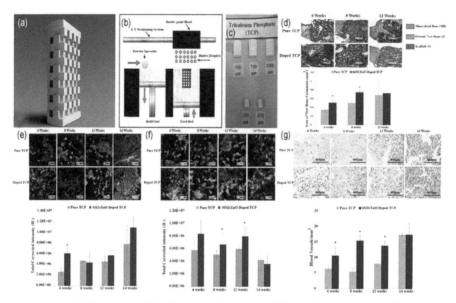

Figure 16. (**a**) CAD design used for 3D printing of porous scaffolds; (**b**) Schematic diagram illustration of 3D printing process (binder jetting); (**c**) Real 3D printed scaffolds, numbers indicating the pore size of scaffolds; (**d**) Staining of implant section via Goldner's trichrome. Gray/brown color shows CaP implants, blue is mineralized implants and osteoid formation can be seen by orange color. Histomorphometry has done on trichrome sections (P < 0.1, where n = 3); (**e**) Confocal micrographs of collagen I formation (green) over the period of 16 weeks; (**f**) Confocal micrographs of osteocalcin (green). While blue color indicates counterstain for cell nuclei; (**g**) Light micrographs showing vWF staining (the dark red spots) [118].

5. Challenges and Future Perspective

Despite all the achievements made in the past in 3D printing of tissue engineering, several challenges still exist. Challenges can be divided into two major categories: (1) 3D printing of biomaterials including live cells and (2) Post-implantation integration and functionality in vivo model. One of the most common problems during manufacturing is nozzle clogging in nozzle-based 3D printing techniques. To overcome nozzle clogging, printing precursor should have proper viscosity and need to be homogenous. Another problem is that the 3D printed constructs need to be adequately stable and mechanically stiff to ensure effective prosthesis. For instance, in hard tissue transplant, the scaffolds elastic modulus should be high enough to sustain its designed porosity and structure to help natural cell growth [128].

3D printed constructs for bone tissue engineering, being eventually implanted in a body, so these constructs also need to support vascularization to deliver sufficient amount of oxygen and nutrition to the cells in vivo to enhance the growth of newly implanted bone [129]. Vascularization plays a pivotal role in a successful bone tissue implant. However, it remains a daunting challenge in bone tissue engineering, particularly, in clinical application of large bone defects. Development of vascularized and clinically applicable bone substitutes with adequate blood supply, capable of inducing angiogenesis and sustaining implant viability remains a critical challenge. Since oxygen is only accessible to those cells through diffusion that are 100–200 μm from blood vessels, bioprinted constructs thicker than 400 μm face oxygenation problem. Therefore, it is a critical task to provide ideal conditions to help vascularization in implanted bone constructs. There is a need for some extensive research to completely understand the mechanism of the biological system of bone. Thus, manufacturing a biomimetic vascularized bone that mimics the native bone can be helpful to overcome these hurdles. Due to the ability of bioprinters utilizing several print-heads loaded with different cell types, introducing vasculature was made possible to a 3D printed construct [130–132].

Recently, nozzle-based 3D-printers enabled the printing of endothelial cells using multiple bioinks for the development of thick vascularized [133,134]. Especially, digital light processing (DLP) based 3D bioprinting can offer extraordinary speed, scalability and resolution for printing complex 3D structures with micrometer resolution [135,136]. For instance, Zhu et al. printed well-designed vascular channels without using perfusion or sacrificial materials by utilizing a rapid microscale continuous optical bioprinter (μCOB). In this method, glycidyl methacrylate-hyaluronic acid (GM-HA) and GelMA-cell laden bioinks were used to create channels and channel adjacent regions. From the outcomes of this study, researchers were able to demonstrate the progressive formation of endothelial network and formation of the lumen-like structures in vivo/vitro model. Anastomosis between the bioprinted endothelial network and circulation was observed with functional blood vessels featuring red blood cells [137]. Moreover, hypoxia is also having an important role in vascularization and bone regeneration process. Hypoxia belongs to the family of Hypoxia-Inducible Transcription Factors (HIFs) [138]. Kuss et al. utilized short-term hypoxic conditions to endorse vascularization in a hybrid 3D printed scaffold of polycaprolactone/hydroxyapatite (PCL/HAp) and stromal vascular fraction (SVF) derived cell laden bioink [139].

Another type of challenge is regulatory hurdles, customized 3D printing technology entails series of difficulties in the regulatory approval field. Though, it is urgent for the managing authorities to establish appropriate laws and regulations to ensure sustainable progress of 3D printing technology. At present, 3D-printed scaffolds and tissues are used for evaluation and screening purposes in animal models.

6. Conclusions

In summary, this review outlined the latest researches on development of 3D printing of bioceramics for bone tissue engineering, current state of the art is also discussed. Extensive amount of research on 3D bioprinting over the past 10 years highlighted its wide range of applications and potentials in bone tissue engineering. Although, plethora of goals have been accomplished in 3D printing of bioceramics, but it is still in its emerging stage.

However, to deal with challenges such as vascularization, and printing related problems, further research on development of bioinks, integration of different 3D bioprinting technologies, improvement of the mechanical properties of existing bioceramics, development of composites with excellent biocompatibility and better understanding of bonding mechanism between bone mineral and collagen are some primary areas of concern that can help to improve the applications of 3D printing in bone tissue engineering.

Recently, a very limited number of bio-printed products have been commercially available. Due to the rapid expansion of this industry in the last few years, it is foreseeable that more bio-printed constructs will ultimately become commercially available to help wide range of patients suffering from different kind of diseases. The technical problems related to clinical requirements and materials selection are mentioned above, multidisciplinary research will be required to tackle those problems and to comprehensively understand the potential of bioprinting in bone tissue engineering.

Author Contributions: Conceptualization, Methodology, Original draft preparation and data curation; M.J.Z. and D.Z. contributed equally; formal analysis, revision and supervision done by D.Z. and Z.Z.

Funding: This work had been financially supported by the Natural Science Foundation of Hebei Province, China (E2018202200).

Conflicts of Interest: The authors declare no conflict of interest.

References

1. Sachs, E.; Cima, M.; Williams, P.; Brancazio, D.; Cornie, J. Three-Dimensional Printing: Rapid Tooling and Prototypes Directly from a CAD Model. *J. Eng. Ind.* **1992**, *114*, 481–488. [CrossRef]
2. Malik, H.H.; Darwood, A.R.J.; Shaunak, S.; Kulatilake, P.; El-Hilly, A.A.; Mulki, O.; Baskaradas, A. Three-dimensional printing in surgery: A review of current surgical applications. *J. Surg. Res.* **2015**, *199*, 512–522. [CrossRef] [PubMed]
3. An, J.; Teoh, J.E.M.; Suntornnond, R.; Chua, C.K. Design and 3D Printing of Scaffolds and Tissues. *Engineering* **2015**, *1*, 261–268. [CrossRef]
4. Ma, H.; Feng, C.; Chang, J.; Wu, C. 3D-printed bioceramic scaffolds: From bone tissue engineering to tumor therapy. *Acta Biomater.* **2018**, *79*, 37–59. [CrossRef]
5. Derby, B. Printing and Prototyping of Tissues and Scaffolds. *Science* **2012**, *338*, 921–926. [CrossRef]
6. Turnbull, G.; Clarke, J.; Picard, F.; Riches, P.; Jia, L.; Han, F.; Li, B.; Shu, W. 3D bioactive composite scaffolds for bone tissue engineering. *Bioact. Mater.* **2018**, *3*, 278–314. [CrossRef]
7. Best, S.M.; Porter, A.E.; Thian, E.S.; Huang, J. Bioceramics: Past, present and for the future. *J. Eur. Ceram. Soc.* **2008**, *28*, 1319–1327. [CrossRef]
8. Blokhuis, T.J.; Arts, J.J.C. Bioactive and osteoinductive bone graft substitutes: Definitions, facts and myths. *Injury* **2011**, *42*, 26–29. [CrossRef]
9. Jones, J.R.; Hench, L.L. Regeneration of trabecular bone using porous ceramics. *Curr. Opin. Solid State Mater. Sci.* **2003**, *7*, 301–307. [CrossRef]
10. Schieker, M.; Seitz, H.; Drosse, I.; Seitz, S.; Mutschler, W.J.E.J.O.T. Biomaterials as Scaffold for Bone Tissue Engineering. *Eur. J. Trauma* **2006**, *32*, 114–124. [CrossRef]
11. Bose, S.; Tarafder, S. Calcium phosphate ceramic systems in growth factor and drug delivery for bone tissue engineering: A review. *Acta Biomater.* **2012**, *8*, 1401–1421. [CrossRef] [PubMed]
12. Wu, S.; Liu, X.; Yeung, K.W.K.; Liu, C.; Yang, X. Biomimetic porous scaffolds for bone tissue engineering. *Mater. Sci. Eng. R Rep.* **2014**, *80*, 1–36. [CrossRef]
13. Brie, J.; Chartier, T.; Chaput, C.; Delage, C.; Pradeau, B.; Caire, F.; Boncoeur, M.-P.; Moreau, J.-J. A new custom made bioceramic implant for the repair of large and complex craniofacial bone defects. *J. Cranio Maxillofac. Surg.* **2013**, *41*, 403–407. [CrossRef] [PubMed]
14. Tang, D.; Tare, R.S.; Yang, L.-Y.; Williams, D.F.; Ou, K.-L.; Oreffo, R.O.C. Biofabrication of bone tissue: Approaches, challenges and translation for bone regeneration. *Biomaterials* **2016**, *83*, 363–382. [CrossRef] [PubMed]
15. Brunello, G.; Sivolella, S.; Meneghello, R.; Ferroni, L.; Gardin, C.; Piattelli, A.; Zavan, B.; Bressan, E. Powder-based 3D printing for bone tissue engineering. *Biotechnol. Adv.* **2016**, *34*, 740–753. [CrossRef] [PubMed]
16. Murphy, S.V.; Atala, A. 3D bioprinting of tissues and organs. *Nat. Biotechnol.* **2014**, *32*, 773. [CrossRef] [PubMed]
17. *ASTM F2792-12a*; Standard Terminology for Additive Manufacturing Technologies, (Withdrawn 2015); ASTM International: West Conshohocken, PA, USA, 2012.
18. Gibson, I.; Rosen, D.; Stucker, B. *Additive Manufacturing Technologies: 3D Printing, Rapid Prototyping, and Direct Digital Manufacturing*; Springer: New York, NY, USA, 2015.
19. Miyanaji, H.; Orth, M.; Akbar, J.M.; Yang, L.J.F.O.M.E. Process development for green part printing using binder jetting additive manufacturing. *Front. Mech. Eng.* **2018**, *13*, 504–512. [CrossRef]
20. Snelling, D.A.; Williams, C.B.; Suchicital, C.T.; Druschitz, A.P. Binder jetting advanced ceramics for metal-ceramic composite structures. *Int. J. Adv. Manuf. Technol.* **2017**, *92*, 531–545. [CrossRef]
21. Cima, L.G.; Cima, M.J. Massachusetts Institute of Technology, Assignee. Preparation of Medical Devices by Solid Free-Form Fabrication. U.S. Patent Application No. 08138345, 13 February 1996.
22. Sachs, E.M.; Haggerty, J.S.; Cima, M.J.; Williams, P.A. Three-Dimensional Printing Techniques. U.S. Patent Application No. 5387380A, 7 February 1995.

23. Lu, K.; Hiser, M.; Wu, W. Effect of particle size on three-dimensional printed mesh structures. *Powder Technol.* **2009**, *192*, 178–183. [CrossRef]

24. Lee, J.-Y.; An, J.; Chua, C.K. Fundamentals and applications of 3D printing for novel materials. *Appl. Mater. Today* **2017**, *7*, 120–133. [CrossRef]

25. Balla, V.K.; Bose, S.; Bandyopadhyay, A. Processing of Bulk Alumina Ceramics Using Laser Engineered Net Shaping. *Int. J. Appl. Ceram. Technol.* **2008**, *5*, 234–242. [CrossRef]

26. Zhai, Y.; Lados, D.A.; Brown, E.J.; Vigilante, G.N. Fatigue crack growth behavior and microstructural mechanisms in Ti-6Al-4V manufactured by laser engineered net shaping. *Int. J. Fatigue* **2016**, *93*, 51–63. [CrossRef]

27. Schiele, N.; Chrisey, D.; Corr, D. Gelatin-Based Laser Direct-Write Technique for the Precise Spatial Patterning of Cells. *Tissue Eng. Part C Methods* **2011**, *17*, 289–298. [CrossRef] [PubMed]

28. Ozbolat, I.T.; Yu, Y. Bioprinting Toward Organ Fabrication: Challenges and Future Trends. *IEEE Trans. Biomed. Eng.* **2013**, *60*, 691–699. [CrossRef] [PubMed]

29. Ghazanfari, A.; Li, W.; Leu, M.; Hilmas, G. A Novel Extrusion-Based Additive Manufacturing Process for Ceramic Parts. In Proceedings of the 27th Annual International Solid Freeform Fabrication Symposium, Austin, TX, USA, 8–10 August 2016; pp. 1509–1529.

30. Deckers, J.; Vleugels, J.; Kruth, J.P. Additive Manufacturing of Ceramics: A Review. *J. Ceram. Sci. Technol.* **2014**, *5*, 245–260.

31. Bose, S.; Suguira, S.; Bandyopadhyay, A. Processing of controlled porosity ceramic structures via fused deposition. *Scr. Mater.* **1999**, *41*, 1009–1014. [CrossRef]

32. Ozbolat, I.T.; Hospodiuk, M. Current advances and future perspectives in extrusion-based bioprinting. *Biomaterials* **2016**, *76*, 321–343. [CrossRef]

33. Ji, S.; Guvendiren, M. Recent Advances in Bioink Design for 3D Bioprinting of Tissues and Organs. *Front. Bioeng. Biotechnol.* **2017**, *5*, 23. [CrossRef]

34. Derby, B. Additive Manufacture of Ceramics Components by Inkjet Printing. *Engineering* **2015**, *1*, 113–123. [CrossRef]

35. Sing, S.L.; Yeong, W.Y.; Wiria, F.E.; Tay, B.Y.; Zhao, Z.; Zhao, L.; Tian, Z.; Yang, S. Direct selective laser sintering and melting of ceramics: A review. *Rapid Prototyp. J.* **2017**, *23*, 611–623. [CrossRef]

36. Qian, B.; Shen, Z. Laser sintering of ceramics. *J. Asian Ceram. Soc.* **2013**, *1*, 315–321. [CrossRef]

37. Kruth, J.P.; Mercelis, P.; Van Vaerenbergh, J.; Froyen, L.; Rombouts, M. Binding mechanisms in selective laser sintering and selective laser melting. *Rapid Prototyp. J.* **2005**, *11*, 26–36. [CrossRef]

38. Bertrand, P.; Bayle, F.; Combe, C.; Goeuriot, P.; Smurov, I. Ceramic components manufacturing by selective laser sintering. *Appl. Surf. Sci.* **2007**, *254*, 989–992. [CrossRef]

39. Mazzoli, A. Selective laser sintering in biomedical engineering. *Med. Biol. Eng. Comput.* **2013**, *51*, 245–256. [CrossRef] [PubMed]

40. Haeri, S. Optimisation of blade type spreaders for powder bed preparation in Additive Manufacturing using DEM simulations. *Powder Technol.* **2017**, *321*, 94–104. [CrossRef]

41. Chartrain, N.A.; Williams, C.B.; Whittington, A.R. A review on fabricating tissue scaffolds using vat photopolymerization. *Acta Biomater.* **2018**, *74*, 90–111. [CrossRef] [PubMed]

42. Melchels, F.P.W.; Feijen, J.; Grijpma, D.W. A review on stereolithography and its applications in biomedical engineering. *Biomaterials* **2010**, *31*, 6121–6130. [CrossRef]

43. Stevens, A.G.; Oliver, C.R.; Kirchmeyer, M.; Wu, J.; Chin, L.; Polsen, E.S.; Archer, C.; Boyle, C.; Garber, J.; Hart, A.J. Conformal Robotic Stereolithography. *3D Print. Addit. Manuf.* **2016**, *3*, 226–235. [CrossRef]

44. Li, X.; Li, D.; Lu, B.; Wang, C. Fabrication of bioceramic scaffolds with pre-designed internal architecture by gel casting and indirect stereolithography techniques. *J. Porous Mater.* **2008**, *15*, 667–671. [CrossRef]

45. Wu, H.; Liu, W.; He, R.; Wu, Z.; Jiang, Q.; Song, X.; Chen, Y.; Cheng, L.; Wu, S. Fabrication of dense zirconia-toughened alumina ceramics through a stereolithography-based additive manufacturing. *Ceram. Int.* **2017**, *43*, 968–972. [CrossRef]

46. Islam, M.N.; Gomer, H.; Sacks, S. Comparison of dimensional accuracies of stereolithography and powder binder printing. *Int. J. Adv. Manuf. Technol.* **2017**, *88*, 3077–3087. [CrossRef]

47. He, L.; Song, X.J.J. Supportability of a High-Yield-Stress Slurry in a New Stereolithography-Based Ceramic Fabrication Process. *JOM* **2018**, *70*, 407–412. [CrossRef]

48. Hench, L.L. Bioceramics and the origin of life. *J. Biomed. Mater. Res.* **1989**, *23*, 685–703. [CrossRef] [PubMed]

49. Habraken, W.; Habibovic, P.; Epple, M.; Bohner, M. Calcium phosphates in biomedical applications: Materials for the future? *Mater. Today* **2016**, *19*, 69–87. [CrossRef]

50. Oonishi, H. Orthopaedic applications of hydroxyapatite. *Biomaterials* **1991**, *12*, 171–178. [CrossRef]

51. Petit, R. The use of hydroxyapatite in orthopaedic surgery: A ten-year review. *Eur. J. Orthop. Surg. Traumatol.* **1999**, *9*, 71–74. [CrossRef]

52. Zeng, Y.; Yan, Y.; Yan, H.; Liu, C.; Li, P.; Dong, P.; Zhao, Y.; Chen, J. 3D printing of hydroxyapatite scaffolds with good mechanical and biocompatible properties by digital light processing. *J. Mater. Sci.* **2018**, *53*, 6291–6301. [CrossRef]

53. Szcześ, A.; Hołysz, L.; Chibowski, E. Synthesis of hydroxyapatite for biomedical applications. *Adv. Colloid Interface Sci.* **2017**, *249*, 321–330. [CrossRef]

54. Zhou, H.; Lee, J. Nanoscale hydroxyapatite particles for bone tissue engineering. *Acta Biomater.* **2011**, *7*, 2769–2781. [CrossRef]

55. Bouler, J.M.; Pilet, P.; Gauthier, O.; Verron, E. Biphasic calcium phosphate ceramics for bone reconstruction: A review of biological response. *Acta Biomater.* **2017**, *53*, 1–12. [CrossRef]

56. Busuttil Naudi, K.; Ayoub, A.; McMahon, J.; Di Silvio, L.; Lappin, D.; Hunter, K.D.; Barbenel, J. Mandibular reconstruction in the rabbit using beta-tricalcium phosphate (β-TCP) scaffolding and recombinant bone morphogenetic protein 7 (rhBMP-7)—Histological, radiographic and mechanical evaluations. *J. Cranio Maxillofac. Surg.* **2012**, *40*, 461–469. [CrossRef] [PubMed]

57. Ryu, H.-S.; Youn, H.-J.; Sun Hong, K.; Chang, B.-S.; Lee, C.-K.; Chung, S.-S. An improvement in sintering property of β-tricalcium phosphate by adition of calcium pyrophosphate. *Biomaterials* **2002**, *23*, 909–914. [CrossRef]

58. Brazete, D.; Torres, P.M.C.; Abrantes, J.C.C.; Ferreira, J.M.F. Influence of the Ca/P ratio and cooling rate on the allotropic α β-tricalcium phosphate phase transformations. *Ceram. Int.* **2018**, *44*, 8249–8256.

59. Fernandes, J.S.; Gentile, P.; Pires, R.A.; Reis, R.L.; Hatton, P.V. Multifunctional bioactive glass and glass-ceramic biomaterials with antibacterial properties for repair and regeneration of bone tissue. *Acta Biomater.* **2017**, *59*, 2–11. [CrossRef]

60. Baino, F.; Fiorilli, S.; Vitale-Brovarone, C. Bioactive glass-based materials with hierarchical porosity for medical applications: Review of recent advances. *Acta Biomater.* **2016**, *42*, 18–32. [CrossRef]

61. Rahaman, M.N.; Day, D.E.; Sonny Bal, B.; Fu, Q.; Jung, S.B.; Bonewald, L.F.; Tomsia, A.P. Bioactive glass in tissue engineering. *Acta Biomater.* **2011**, *7*, 2355–2373. [CrossRef]

62. Liu, H.; Yazici, H.; Ergun, C.; Webster, T.J.; Bermek, H. An in vitro evaluation of the Ca/P ratio for the cytocompatibility of nano-to-micron particulate calcium phosphates for bone regeneration. *Acta Biomater.* **2008**, *4*, 1472–1479. [CrossRef]

63. Parent, M.; Baradari, H.; Champion, E.; Damia, C.; Viana-Trecant, M. Design of calcium phosphate ceramics for drug delivery applications in bone diseases: A review of the parameters affecting the loading and release of the therapeutic substance. *J. Control. Release* **2017**, *252*, 1–17. [CrossRef]

64. Ahmad, O.; Soodeh, A. Application of Bioceramics in Orthopedics and Bone Tissue Engineering. Available online: https://www.researchgate.net/publication/321939283_Application_of_Bioceramics_in_Orthopedics_and_Bone_Tissue_Engineering (accessed on 11 October 2019).

65. Hench, L.L. Bioceramics: From Concept to Clinic. *J. Am. Ceram. Soc.* **1991**, *74*, 1487–1510. [CrossRef]

66. Chevalier, J.; Gremillard, L. Ceramics for medical applications: A picture for the next 20 years. *J. Eur. Ceram. Soc.* **2009**, *29*, 1245–1255. [CrossRef]

67. Dorozhkin, S.V. Calcium orthophosphate bioceramics. *Ceram. Int.* **2015**, *41*, 13913–13966. [CrossRef]

68. Oonishi, H.; Oonishi, H.; Ohashi, H.; Kawahara, I.; Hanaoka, Y.; Iwata, R.; Hench, L.L. Clinical Applications of Hydroxyapatite in Orthopedics. In *Advances in Calcium Phosphate Biomaterials*; Ben-Nissan, B., Ed.; Springer Berlin Heidelberg: Berlin/Heidelberg, Germany, 2014; pp. 19–49. [CrossRef]

69. Asri, R.I.M.; Harun, W.S.W.; Hassan, M.A.; Ghani, S.A.C.; Buyong, Z. A review of hydroxyapatite-based coating techniques: Sol-gel and electrochemical depositions on biocompatible metals. *J. Mech. Behav. Biomed. Mater.* **2016**, *57*, 95–108. [CrossRef] [PubMed]

70. Cox, S.C.; Thornby, J.A.; Gibbons, G.J.; Williams, M.A.; Mallick, K.K. 3D printing of porous hydroxyapatite scaffolds intended for use in bone tissue engineering applications. *Mater. Sci. Eng. C* **2015**, *47*, 237–247. [CrossRef] [PubMed]

71. Ayoub, G.; Veljovic, D.; Zebic, M.L.; Miletic, V.; Palcevskis, E.; Petrovic, R.; Janackovic, D. Composite nanostructured hydroxyapatite/yttrium stabilized zirconia dental inserts—The processing and application as dentin substitutes. *Ceram. Int.* **2018**, *44*, 18200–18208. [CrossRef]

72. Hung, K.-Y.; Lo, S.-C.; Shih, C.-S.; Yang, Y.-C.; Feng, H.-P.; Lin, Y.-C. Titanium surface modified by hydroxyapatite coating for dental implants. *Surf. Coat. Technol.* **2013**, *231*, 337–345. [CrossRef]

73. Ciobanu, G.; Harja, M. Cerium-doped hydroxyapatite/collagen coatings on titanium for bone implants. *Ceram. Int.* **2019**, *45*, 2852–2857. [CrossRef]

74. Shi, P.; Liu, M.; Fan, F.; Yu, C.; Lu, W.; Du, M. Characterization of natural hydroxyapatite originated from fish bone and its biocompatibility with osteoblasts. *Mater. Sci. Eng. C* **2018**, *90*, 706–712. [CrossRef]

75. Carfi Pavia, F.; Conoscenti, G.; Greco, S.; La Carrubba, V.; Ghersi, G.; Brucato, V. Preparation, characterization and in vitro test of composites poly-lactic acid/hydroxyapatite scaffolds for bone tissue engineering. *Int. J. Biol. Macromol.* **2018**, *119*, 945–953. [CrossRef]

76. He, X.; Fan, X.; Feng, W.; Chen, Y.; Guo, T.; Wang, F.; Liu, J.; Tang, K. Incorporation of microfibrillated cellulose into collagen-hydroxyapatite scaffold for bone tissue engineering. *Int. J. Biol. Macromol.* **2018**, *115*, 385–392. [CrossRef]

77. Sposito Corcione, C.; Gervaso, F.; Scalera, F.; Padmanabhan, S.K.; Madaghiele, M.; Montagna, F.; Sannino, A.; Licciulli, A.; Maffezzoli, A. Highly loaded hydroxyapatite microsphere/ PLA porous scaffolds obtained by fused deposition modelling. *Ceram. Int.* **2019**, *45*, 2803–2810. [CrossRef]

78. Barry, J.J.A.; Evseev, A.V.; Markov, M.A.; Upton, C.E.; Scotchford, C.A.; Popov, V.K.; Howdle, S.M. In vitro study of hydroxyapatite-based photocurable polymer composites prepared by laser stereolithography and supercritical fluid extraction. *Acta Biomater.* **2008**, *4*, 1603–1610. [CrossRef] [PubMed]

79. Li, Q.; Lei, X.; Wang, X.; Cai, Z.; Lyu, P.; Zhang, G. Hydroxyapatite/Collagen Three-Dimensional Printed Scaffolds and Their Osteogenic Effects on Human Bone Marrow-Derived Mesenchymal Stem Cells. *Tissue Eng. Part A* **2019**, *25*, 1261–1271. [CrossRef] [PubMed]

80. Woesz, A.; Rumpler, M.; Stampfl, J.; Varga, F.; Fratzl-Zelman, N.; Roschger, P.; Klaushofer, K.; Fratzl, P. Towards bone replacement materials from calcium phosphates via rapid prototyping and ceramic gelcasting. *Mater. Sci. Eng. C* **2005**, *25*, 181–186. [CrossRef]

81. Saiz, E.; Gremillard, L.; Menendez, G.; Miranda, P.; Gryn, K.; Tomsia, A.P. Preparation of porous hydroxyapatite scaffolds. *Mater. Sci. Eng. C* **2007**, *27*, 546–550. [CrossRef]

82. Virginie, K.; Fabien, G.; Isabelle, A.; Bertrand, G.; Sylvain, M.; Joëlle, A.; Jean-Christophe, F.; Sylvain, C. In vivo bioprinting for computer- and robotic-assisted medical intervention: preliminary study in mice. *Biofabrication* **2010**, *2*, 014101.

83. Tian, Y.; Lu, T.; He, F.; Xu, Y.; Shi, H.; Shi, X.; Zuo, F.; Wu, S.; Ye, J. β-tricalcium phosphate composite ceramics with high compressive strength, enhanced osteogenesis and inhibited osteoclastic activities. *Colloids Surf. B Biointerfaces* **2018**, *167*, 318–327. [CrossRef]

84. Hirakawa, Y.; Manaka, T.; Orita, K.; Ito, Y.; Ichikawa, K.; Nakamura, H. The accelerated effect of recombinant human bone morphogenetic protein 2 delivered by β-tricalcium phosphate on tendon-to-bone repair process in rabbit models. *J. Shoulder Elb. Surg.* **2018**, *27*, 894–902. [CrossRef]

85. Cheng, L.; Duan, X.; Xiang, Z.; Shi, Y.; Lu, X.; Ye, F.; Bu, H. Ectopic bone formation cannot occur by hydroxyapatite/β-tricalcium phosphate bioceramics in green fluorescent protein chimeric mice. *Appl. Surf. Sci.* **2012**, *262*, 200–206. [CrossRef]

86. Stähli, C.; Bohner, M.; Bashoor-Zadeh, M.; Doebelin, N.; Baroud, G. Aqueous impregnation of porous β-tricalcium phosphate scaffolds. *Acta Biomater.* **2010**, *6*, 2760–2772. [CrossRef]

87. Horch, H.H.; Sader, R.; Pautke, C.; Neff, A.; Deppe, H.; Kolk, A. Synthetic, pure-phase beta-tricalcium phosphate ceramic granules for bone regeneration in the reconstructive surgery of the jaws. *Int. J. Oral Maxillofac. Surg.* **2006**, *35*, 708–713. [CrossRef]

88. Zerbo, I.R.; Bronckers, A.L.J.J.; De Lange, G.L.; Burger, E.H.; Van Beek, G.J. Histology of human alveolar bone regeneration with a porous tricalcium phosphate. *Clin. Oral Implant. Res.* **2001**, *12*, 379–384. [CrossRef]

89. Li, B.; Liu, Z.; Yang, J.; Yi, Z.; Xiao, W.; Liu, X.; Yang, X.; Xu, W.; Liao, X. Preparation of bioactive β-tricalcium phosphate microspheres as bone graft substitute materials. *Mater. Sci. Eng. C* **2017**, *70*, 1200–1205. [CrossRef] [PubMed]

90. Cao, Y.; Xiao, L.; Cao, Y.; Nanda, A.; Xu, C.; Ye, Q. 3D printed β-TCP scaffold with sphingosine 1-phosphate coating promotes osteogenesis and inhibits inflammation. *Biochem. Biophys. Res. Commun.* **2019**, *512*, 889–895. [CrossRef] [PubMed]

91. Bian, W.; Li, D.; Lian, Q.; Li, X.; Zhang, W.; Wang, K.; Jin, Z. Fabrication of a bio-inspired beta-Tricalcium phosphate/collagen scaffold based on ceramic stereolithography and gel casting for osteochondral tissue engineering. *Rapid Prototyp. J.* **2012**, *18*, 68–80. [CrossRef]

92. Bose, S.; Banerjee, D.; Robertson, S.; Vahabzadeh, S. Enhanced In Vivo Bone and Blood Vessel Formation by Iron Oxide and Silica Doped 3D Printed Tricalcium Phosphate Scaffolds. *Ann. Biomed. Eng.* **2018**, *46*, 1241–1253. [CrossRef]

93. Tarafder, S.; Balla, V.K.; Davies, N.M.; Bandyopadhyay, A.; Bose, S. Microwave-sintered 3D printed tricalcium phosphate scaffolds for bone tissue engineering. *J. Tissue Eng. Regen. Med.* **2013**, *7*, 631–641. [CrossRef]

94. Giannoudis, P.V.; Dinopoulos, H.; Tsiridis, E. Bone substitutes: An update. *Injury* **2005**, *36*, 20–27. [CrossRef]

95. Välimäki, V.-V.; Aro, H. Molecular basis for action of bioactive glasses as bone graft substitute. *Scand. J. Surg.* **2006**, *95*, 95–102. [CrossRef]

96. Eqtesadi, S.; Motealleh, A.; Miranda, P.; Pajares, A.; Lemos, A.; Ferreira, J.M.F. Robocasting of 45S5 bioactive glass scaffolds for bone tissue engineering. *J. Eur. Ceram. Soc.* **2014**, *34*, 107–118. [CrossRef]

97. Xynos, I.D.; Edgar, A.J.; Buttery, L.D.K.; Hench, L.L.; Polak, J.M. Gene-expression profiling of human osteoblasts following treatment with the ionic products of Bioglass® 45S5 dissolution. *J. Biomed. Mater. Res.* **2001**, *55*, 151–157. [CrossRef]

98. Wu, C.; Luo, Y.; Cuniberti, G.; Xiao, Y.; Gelinsky, M. Three-dimensional printing of hierarchical and tough mesoporous bioactive glass scaffolds with a controllable pore architecture, excellent mechanical strength and mineralization ability. *Acta Biomater.* **2011**, *7*, 2644–2650. [CrossRef] [PubMed]

99. Pei, P.; Tian, Z.; Zhu, Y. 3D printed mesoporous bioactive glass/metal-organic framework scaffolds with antitubercular drug delivery. *Microporous Mesoporous Mater.* **2018**, *272*, 24–30. [CrossRef]

100. Baino, F. Bioactive glasses—When glass science and technology meet regenerative medicine. *Ceram. Int.* **2018**, *44*, 14953–14966. [CrossRef]

101. Hsu, F.-Y.; Hsu, H.-W.; Chang, Y.-H.; Yu, J.-L.; Rau, L.-R.; Tsai, S.-W. Macroporous microbeads containing apatite-modified mesoporous bioactive glass nanofibres for bone tissue engineering applications. *Mater. Sci. Eng. C* **2018**, *89*, 346–354. [CrossRef]

102. Nommeots-Nomm, A.; Lee, P.D.; Jones, J.R. Direct ink writing of highly bioactive glasses. *J. Eur. Ceram. Soc.* **2018**, *38*, 837–844. [CrossRef]

103. Padilla, S.; Sánchez-Salcedo, S.; Vallet-Regí, M. Bioactive glass as precursor of designed-architecture scaffolds for tissue engineering. *J. Biomed. Mater. Res. Part A* **2007**, *81*, 224–232. [CrossRef]

104. Westhauser, F.; Weis, C.; Prokscha, M.; Bittrich, L.A.; Li, W.; Xiao, K.; Kneser, U.; Kauczor, H.-U.; Schmidmaier, G.; Boccaccini, A.R.; et al. Three-dimensional polymer coated 45S5-type bioactive glass scaffolds seeded with human mesenchymal stem cells show bone formation in vivo. *J. Mater. Sci. Mater. Med.* **2016**, *27*, 119. [CrossRef]

105. Liu, Z.; Liang, H.; Shi, T.; Xie, D.; Chen, R.; Han, X.; Shen, L.; Wang, C.; Tian, Z. Additive manufacturing of hydroxyapatite bone scaffolds via digital light processing and in vitro compatibility. *Ceram. Int.* **2019**, *45*, 11079–11086. [CrossRef]

106. Wang, Y.; Wang, K.; Li, X.; Wei, Q.; Chai, W.; Wang, S.; Che, Y.; Lu, T.; Zhang, B. 3D fabrication and characterization of phosphoric acid scaffold with a HA/beta-TCP weight ratio of 60:40 for bone tissue engineering applications. *PLoS ONE* **2017**, *12*, e0174870.

107. Tarafder, S.; Bose, S. Polycaprolactone-Coated 3D Printed Tricalcium Phosphate Scaffolds for Bone Tissue Engineering: In Vitro Alendronate Release Behavior and Local Delivery Effect on In Vivo Osteogenesis. *ACS Appl. Mater. Interfaces* **2014**, *6*, 9955–9965. [CrossRef]

108. Luo, G.; Ma, Y.; Cui, X.; Jiang, L.; Wu, M.; Hu, Y.; Luo, Y.; Pan, H.; Ruan, C. 13-93 bioactive glass/alginate composite scaffolds 3D printed under mild conditions for bone regeneration. *RSC Adv.* **2017**, *7*, 11880–11889. [CrossRef]

109. Dimitriou, R.; Jones, E.; McGonagle, D.; Giannoudis, P. Bone regeneration: current concepts and future directions. *BMC Med.* **2011**, *9*, 66. [CrossRef] [PubMed]

110. Pilia, M.; Guda, T.; Appleford, M. Development of Composite Scaffolds for Load-Bearing Segmental Bone Defects. *BioMed Res. Int.* **2013**, *2013*, 458253. [CrossRef] [PubMed]

111. Reichert, J.C.; Wullschleger, M.E.; Cipitria, A.; Lienau, J.; Cheng, T.K.; Schütz, M.A.; Duda, G.N.; Nöth, U.; Eulert, J.; Hutmacher, D. Custom-made composite scaffolds for segmental defect repair in long bones. *Int. Orthop.* **2011**, *35*, 1229–1236. [CrossRef]

112. Wagoner Johnson, A.J.; Herschler, B.A. A review of the mechanical behavior of CaP and CaP/polymer composites for applications in bone replacement and repair. *Acta Biomater.* **2011**, *7*, 16–30. [CrossRef]

113. Chengtie, W.; Jiang, C. A review of bioactive silicate ceramics. *Biomed. Mater.* **2013**, *8*, 032001.

114. Fu, Q.; Saiz, E.; Rahaman, M.N.; Tomsia, A.P. Bioactive glass scaffolds for bone tissue engineering: state of the art and future perspectives. *Mater. Sci. Eng. C* **2011**, *31*, 1245–1256. [CrossRef]

115. Rezwan, K.; Chen, Q.Z.; Blaker, J.J.; Boccaccini, A.R. Biodegradable and bioactive porous polymer/inorganic composite scaffolds for bone tissue engineering. *Biomaterials* **2006**, *27*, 3413–3431. [CrossRef]

116. Vorndran, E.; Klarner, M.; Klammert, U.; Grover, L.M.; Patel, S.; Barralet, J.E.; Gbureck, U. 3D Powder Printing of β-Tricalcium Phosphate Ceramics Using Different Strategies. *Adv. Eng. Mater.* **2008**, *10*, 67–71. [CrossRef]

117. Detsch, R.; Schaefer, S.; Deisinger, U.; Ziegler, G.; Seitz, H.; Leukers, B. In vitro -Osteoclastic Activity Studies on Surfaces of 3D Printed Calcium Phosphate Scaffolds. *J. Biomater. Appl.* **2011**, *26*, 359–380. [CrossRef]

118. Fielding, G.; Bose, S. SiO$_2$ and ZnO dopants in three-dimensionally printed tricalcium phosphate bone tissue engineering scaffolds enhance osteogenesis and angiogenesis in vivo. *Acta Biomater.* **2013**, *9*, 9137–9148. [CrossRef] [PubMed]

119. Wang, Y.; Li, X.; Wei, Q.; Yang, M.; Wei, S. Study on the Mechanical Properties of Three-Dimensional Directly Binding Hydroxyapatite Powder. *Cell Biochem. Biophys.* **2015**, *72*, 289–295. [CrossRef] [PubMed]

120. Miguel, C.; Marta, D.; Elke, V.; Uwe, G.; Paulo, F.; Inês, P.; Barbara, G.; Henrique, A.; Eduardo, P.; Jorge, R. Application of a 3D printed customized implant for canine cruciate ligament treatment by tibial tuberosity advancement. *Biofabrication* **2014**, *6*, 025005.

121. Butscher, A.; Bohner, M.; Doebelin, N.; Hofmann, S.; Müller, R. New depowdering-friendly designs for three-dimensional printing of calcium phosphate bone substitutes. *Acta Biomater.* **2013**, *9*, 9149–9158. [CrossRef] [PubMed]

122. Chumnanklang, R.; Panyathanmaporn, T.; Sitthiseripratip, K.; Suwanprateeb, J. 3D printing of hydroxyapatite: Effect of binder concentration in pre-coated particle on part strength. *Mater. Sci. Eng. C* **2007**, *27*, 914–921. [CrossRef]

123. Suwanprateeb, J.; Sanngam, R.; Suvannapruk, W.; Panyathanmaporn, T. Mechanical and in vitro performance of apatite—Wollastonite glass ceramic reinforced hydroxyapatite composite fabricated by 3D-printing. *J. Mater. Sci. Mater. Med.* **2009**, *20*, 1281. [CrossRef]

124. Hwa, L.C.; Rajoo, S.; Noor, A.M.; Ahmad, N.; Uday, M.B. Recent advances in 3D printing of porous ceramics: A review. *Curr. Opin. Solid State Mater. Sci.* **2017**, *21*, 323–347. [CrossRef]

125. Roohani-Esfahani, S.-I.; Newman, P.; Zreiqat, H. Design and Fabrication of 3D printed Scaffolds with a Mechanical Strength Comparable to Cortical Bone to Repair Large Bone Defects. *Sci. Rep.* **2016**, *6*, 19468. [CrossRef]

126. Fierz, F.C.; Beckmann, F.; Huser, M.; Irsen, S.H.; Leukers, B.; Witte, F.; Degistirici, Ö.; Andronache, A.; Thie, M.; Müller, B. The morphology of anisotropic 3D-printed hydroxyapatite scaffolds. *Biomaterials* **2008**, *29*, 3799–3806. [CrossRef]

127. Dellinger, J.G.; Eurell, J.A.C.; Jamison, R.D. Bone response to 3D periodic hydroxyapatite scaffolds with and without tailored microporosity to deliver bone morphogenetic protein 2. *J. Biomed. Mater. Res.* **2006**, *76*, 366–376. [CrossRef]

128. Hollinger, J.O.; Brekke, J.; Gruskin, E.; Lee, D. Role of Bone Substitutes. *Clin. Orthop. Relat. Res.* **1996**, *324*, 55–65. [CrossRef] [PubMed]

129. Kaully, T.; Kaufman-Francis, K.; Lesman, A.; Levenberg, S. Vascularization—The Conduit to Viable Engineered Tissues. *Tissue Eng. Part B Rev.* **2009**, *15*, 159–169. [CrossRef] [PubMed]

130. Shahabipour, F.; Ashammakhi, N.; Oskuee, R.K.; Bonakdar, S.; Hoffman, T.; Shokrgozar, M.A.; Khademhosseini, A. Key components of engineering vascularized 3-dimensional bioprinted bone constructs. *Transl. Res.* **2019**. [CrossRef] [PubMed]

131. Jammalamadaka, U.; Tappa, K. Recent Advances in Biomaterials for 3D Printing and Tissue Engineering. *J. Funct. Biomater.* **2018**, *9*, 22. [CrossRef] [PubMed]

Materials **2019**, *12*, 3361

132. Tappa, K.; Jammalamadaka, U. Novel Biomaterials Used in Medical 3D Printing Techniques. *J. Funct. Biomater.* **2018**, *9*, 17. [CrossRef]

133. Kolesky, D.B.; Homan, K.A.; Skylar-Scott, M.A.; Lewis, J.A. Three-dimensional bioprinting of thick vascularized tissues. *Proc. Natl. Acad. Sci. USA* **2016**, *113*, 3179–3184. [CrossRef]

134. Bertassoni, L.; Cecconi, M.; Manoharan, V.; Nikkhah, M.; Hjortnaes, J.; Cristino, A.; Barabaschi, G.; Demarchi, D.; Dokmeci, M.; Yang, Y.; et al. Hydrogel Bioprinted Microchannel Networks for Vascularization of Tissue Engineering Constructs. *Lab Chip* **2014**, *14*, 2202–2211. [CrossRef]

135. Zhang, A.P.; Qu, X.; Soman, P.; Hribar, K.C.; Lee, J.W.; Chen, S.; He, S. Rapid Fabrication of Complex 3D Extracellular Microenvironments by Dynamic Optical Projection Stereolithography. *Adv. Mater.* **2012**, *24*, 4266–4270. [CrossRef]

136. Tumbleston, J.; Shirvanyants, D.; Ermoshkin, N.; Janusziewicz, R.; Johnson, A.; Kelly, D.; Chen, K.; Pinschmidt, R.; Rolland, J.; Ermoshkin, A.; et al. Additive manufacturing. Continuous liquid interface production of 3D objects. *Science* **2015**, *347*, 1349–1352. [CrossRef]

137. Zhu, W.; Qu, X.; Zhu, J.; Ma, X.; Patel, S.; Liu, J.; Wang, P.; Lai, C.S.E.; Gou, M.; Xu, Y.; et al. Direct 3D bioprinting of prevascularized tissue constructs with complex microarchitecture. *Biomaterials* **2017**, *124*, 106–115. [CrossRef]

138. Araldi, E.; Schipani, E. Hypoxia, HIFs and bone development. *Bone* **2010**, *47*, 190–196. [CrossRef] [PubMed]

139. Kuss, M.A.; Harms, R.; Wu, S.; Wang, Y.; Untrauer, J.B.; Carlson, M.A.; Duan, B. Short-term hypoxic preconditioning promotes prevascularization in 3D bioprinted bone constructs with stromal vascular fraction derived cells. *RSC Adv.* **2017**, *7*, 29312–29320. [CrossRef] [PubMed]

 *materials*

Review

Materials for Hip Prostheses: A Review of Wear and Loading Considerations

Massimiliano Merola and Saverio Affatato *

Laboratorio di Tecnologia Medica, IRCCS—Istituto Ortopedico Rizzoli, Via di Barbiano, 1/10 40136 Bologna, Italy; merola@tecno.ior.it
* Correspondence: affatato@tecno.ior.it; Tel.: +39-051-6366864; Fax: +39-051-6366863

Received: 11 January 2019; Accepted: 31 January 2019; Published: 5 February 2019

Abstract: Replacement surgery of hip joint consists of the substitution of the joint with an implant able to recreate the articulation functionality. This article aims to review the current state of the art of the biomaterials used for hip implants. Hip implants can be realized with different combination of materials, such as metals, ceramics and polymers. In this review, we analyze, from international literature, the specific characteristics required for biomaterials used in hip joint arthroplasty, i.e., being biocompatible, resisting heavy stress, opposing low frictional forces to sliding and having a low wear rate. A commentary on the evolution and actual existing hip prostheses is proposed. We analyzed the scientific literature, collecting information on the material behavior and the human-body response to it. Particular attention has been given to the tribological behavior of the biomaterials, as friction and wear have been key aspects to improve as hip implants evolve. After more than 50 years of evolution, in term of designs and materials, the actual wear rate of the most common implants is low, allowing us to sensibly reduce the risk related to the widespread debris distribution in the human body.

Keywords: biomaterials; ceramic; friction; hip; implants; polyethylene; prosthesis; simulator; wear

1. Introduction

The hip is one of the most important joints that support our body, having the task of joining the femurs with the pelvis. The smooth and spherical head of the femur fits perfectly into the natural seat of the acetabulum, which is a cup-shaped cavity; the whole joint is wrapped in very resistant ligaments that make the joint stable. The hip joint is subjected to high daily stresses, having to bear the weight of the upper part of the body. Thus, especially with advancing age, these stresses can jeopardize its functioning.

Osteoarthritis of the hip is one of the most widespread alterations of the hip: it is a condition that causes intense pain due to a stiffening of the joint itself. The surface of the femoral head, due to arthritis, can undergo some alterations, becoming porous and causing damage to the entire joint complex. Osteoarthritis of the hip, as a degenerative pathology, involves irreversible damage due to which in many cases it is necessary to resort to the substitution of the compromised joint with an artificial one. A hip prosthesis is an artificial joint designed to perform the same functions as the natural one and which is surgically implanted. The surgical operation is referred to as Total Hip Arthroplasty (THA).

This paper aims to exhaustively review the state of the art of the biomaterials used as hip joint medical devices. More in depth, our review focuses on advantages, disadvantages and future perspectives regarding the use of biomaterials: polymers, metals, ceramics, and composites. This perspective may provide a clearer insight into how biomaterials research sets up the basis for the design of innovative devices for improved solutions to orthopaedic clinical problems.

1.1. History

Since its first application, the development of design and materials of hip prosthesis continuously progressed. Its development is one of the most challenging issues of the century in the field of implant technology [1]. Several materials were used for this scope: glass, polymers, metal alloys, ceramics, composites, etc., trying to combine biocompatibility and fatigue resistance, stiffness, toughness, withstanding static and dynamic loads, and high resistance to mechanical and chemical wear [2,3]. All these biomaterials were developed with the aim to improve the patient's quality life, avoiding repeated surgery. First attempts at hip surgery date back to 1750, in England, willing to heal arthritis cases [4]. In 1840, the first idea of healing the hip was to replace it with a prosthesis [5]. This procedure was limited to resurfacing or replacing the acetabular part of the femoral head. To do so a wooden block was installed between the damaged terminal parts of the hip articulation. Due to wear particles released into the body, this procedure ended up being disastrous. Biological elements were therefore applied to solve the compatibility issue: skin, muscle tissue, pig bladder and gold foil [6]. Only several decades later were used different artificial materials, such as rubber, zinc, glass, wax and silver plates [4]. In 1880, Prof. Themistocles Glück implanted, for the first time, an ivory ball and socket prosthesis fixed to the bone by screws [7]. Later on, finding that human body could not accept large quantities of external material, he experimented with a mixture of plaster of Paris in combination with powder pumice and resin.

Different materials were also introduced: in 1919, Delbet used rubber to replace a femoral head, whereas Hey-Groves used ivory nail in 1922 to simulate the articular surface of the femoral head [5]. In 1925, Marius Smith-Petersen introduced the first glass and bakelite femoral cup, defining the mold arthroplasty technique, that consisted of a hollow hemisphere adapted over the femoral head [8]. In 1938, Philip Wiles performed the first THA, employing a custom-made implant in stainless steel that was fixed to the bone tissue with screws and bolts. In 1950, Austin Moore introduced hemiarthroplasty, a new kind of hip implant, consisting of the replacement of the femoral head and part of the femoral neck using a long-stemmed element. The stem fitted into the femur cavity without cement, substituting around 31 cm of the proximal part of the bone, whereas the ball was placed on the hip acetabulum. This procedure was satisfactory, even though loosening of the implant was still a problem [5]. In Figure 1 are some of the mentioned hip prostheses designs.

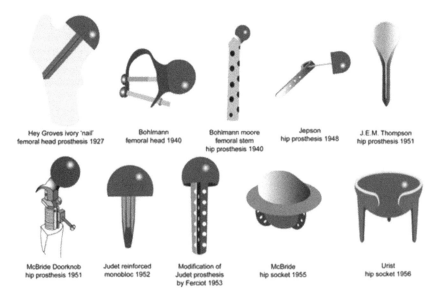

Hey Groves Ivory 'nail' femoral head prosthesis 1927

Bohlmann femoral head 1940

Bohlmann moore femoral stem hip prosthesis 1940

Jepson hip prosthesis 1948

J.E.M. Thompson hip prosthesis 1951

McBride Doorknob hip prosthesis 1951

Judet reinforced monobloc 1952

Modification of Judet prosthesis by Ferciot 1953

McBride hip socket 1955

Urist hip socket 1956

Figure 1. Evolution of the prostheses design.

In 1960, the orthopedic surgeon, San Baw, started performing hip replacements, and in twenty years of work, over 300 ivory hip replacements, with an 88% rate of success [9]. The recognized pioneer of THA, as currently known, is believed to be Sir John Charnley. During the 60's, he defined the concept of Low Friction Arthroplasty (LFA). His first prosthesis was made of a stainless-steel stem, fixed with acrylic cement, and a 22.2-mm diameter head coupled with a polytetrafluoroethylene (PTFE) cup, as shown in Figure 2. PTFE was unsuitable for prosthetic bearing, as it caused wear and tear that leaded to inflammatory reactions. To solve these issues, Sir Charnley adopted other polymer materials, such as high-density polyethylene (HDPE), and ultra-high molecular weight polyethylene (UHMWPE). He also used cement fixation for the acetabular cup [10]. With this combination, the wear effects were reduced, due to the smaller contacting surface and the hard-on-soft coupling. Sir Charnely made many variations to the original design of his LFA, which led to thousands of successful operations.

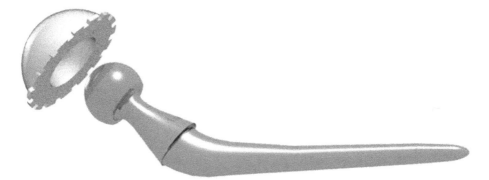

Figure 2. Charnley's first LFA.

1.2. Current Materials

Four main types of bearings are studied and applied in THA: metal-on-polyethylene (MoP), metal-on-metal (MoM), ceramic-on-ceramic (CoC), and ceramic-on-polyethylene (CoP). Recently, hybrid combinations were introduced such as ceramic heads and metallic inserts (CoM) [11,12]. Many factors influence the choice between these types of bearing, such as the implant cost, age and activity level of the patient, complications during surgery, etc. MoM articulations were introduced first in 1950, by McKee and Farrar, leading to unsatisfactory results as two out of three implants were removed after 1 year due to loosening and the third removed for fracture [13–15]. After many improvements of the bearings, they were reintroduced in 1960, when the wear rate ranged from 1 to 5 mm^3 per year (which was roughly 20 times lower than that registered for metal on polyethylene) [16,17]. MoM articulations were used for both total hip replacements and hip resurfacing (HR), which have the advantage of preserving the femoral head and neck, resulting in a less invasive operation and a lower dislocation rate. When, during the 2000s, the issues of metal debris came to light, the MoM replacements were almost stopped completely. In the early middle 2000s, these implants were used in more than one out of five cases in the UK and up to one-third in the US. Today, they are used in less than 1% of the total surgical operations [18]. MoM articulations have been used again in the last two decades, thanks to the appearance of new surface finishing techniques [6] that improve their wear resistance. On the other hand, MoM bearings aim to ensure high wear resistance, good manufacturability and low friction torque. However, even if lower wear volume is associated with such implants, very small particles are produced [19]. The amount of metal ions present in the serum and their potential toxic effects both locally and systemically are a cause for concern [19]. Moreover, polishing wear, promoted by wear debris, produced by the abrasive action of carbides, has been shown in retrieved Co-Cr alloy hip implants [19].

Up to the middle of the 1990s, the most widespread hip implant was MoP couples that worked well in older and less active patients [20]. Two relevant problems were still a concern: aseptic loosening as result of inadequate initial fixation caused by particle-induced osteolysis around the implant and hip dislocation.

In the 80's, when aseptic loosening and osteolysis arose as main issues in metal-on-polymers hip implants, the firsts CoC couples were launched, starting with alumina and zirconia [21–23]. Zirconia ceramics have been introduced for orthopedic implants as a secondary ceramic material along with alumina for several years. Major advantages of ceramics for THA are their hardness, scratch resistance, and the inert nature of debris [24]. These characteristics promote the use of CoC bearings, and the inert nature of the wear debris result in them being the best choice for young patients. On the other hand, their use is expensive, and implants require an excellent surgical insertion to preclude chipping of contact surfaces.

The introduction of an innovative hybrid hard-on-hard bearing ceramic head and metallic insert claimed to reduce ion release and wear particle production and possibly the breakage of the ceramic insert rim [25–27]. In in vitro studies on CoM hip implants [12,28], smaller particles and lower wear have been found.

Nowadays hip joint prostheses are made with metals, ceramics and plastic materials. Most used are titanium alloys, stainless steel, special high-strength alloys, alumina, zirconia, zirconia toughened alumina (ZTA), and UHMWPE. Usually, stems and necks are composed of metals, whereas femoral heads can be both metal and ceramic, and the acetabulum can be made of metals, ceramics or polymers. There are several combinations that can be realized by using these materials with the aim of coupling with the fewest concerns and the highest long-term success odds.

Hereafter, we present an overall evaluation of biomaterials (polymers, metals, ceramics) for THA.

2. Polymers

Polymer materials were the first choice for low friction hip replacements, as proven by Charnely. Highly stable polymeric systems such as PTFE, UHMWPE or polyetheretherketone (PEEK) have been investigated due to their excellent mechanical properties and their high wear resistance. Nevertheless, when implanted, acetabular cups made of polyethylene generate debris that is attacked by the body's immune system [29]. This leads to bone loss, also known as osteolysis; furthermore, since the debris accumulates in the area close to the implant, the bone loss leads to loosening of the implant stem. This results in the needs of a revision, namely, another surgery. Revision for loosening is four times higher than the next leading reason (dislocation at 13.6%) and is more severe in young patients [30].

2.1. PTFE

PTFE has a high thermal stability; it is hydrophobic, stable in most types of chemical environments, and generally considered to be inert in the body [31]. It was used by Charnley in his firsts THA, but exhibited two main drawbacks, which were found only after implantation in 300 patients [32]. The material had a very high wear rate, equal to 0.5 mm per month [33], and PTFE produced voluminous masses of amorphous material due to the vast number of foreign-body giant cells [34]. Furthermore, this debris elicited an intense foreign-body reaction that Charnley verified by injecting two specimens of finely divided PTFE into his own thigh [35].

Charnley tried to use a composite material based on PTFE reinforced with glass fibers (known as Fluorosint), finding poor performance in vivo, despite its fine behavior in vitro. The composite, after one year of implantation, developed a pasty surface that could be easily worn away. Plus, the filler acted abrasively and lapped the metal counter-face. Moreover, this composite material showed a higher rate of infection (20%) and loosening (57%) than the other materials employed [36].

2.2. UHMWPE

Charnley introduced UHMWPE in 1962, urged by the failure of PTFE as a bearing material and sustained by the promising behavior in laboratory tests [37]. The polymer is characterized by its excellent wear resistance, low friction and high impact strength. It is created by the polymerization of ethylene, and it is one of the simplest polymers. Its chemical formula is $(-C_2H_4-)_n$, where n is the degree of polymerization, being the number of repeating units along the chain. The average degree of n is a minimum of 36,000 [38], having a molecular weight of at least 1 million g/mole as defined by the standard [39].

During the 1980s and early 1990s, aseptic loosening and osteolysis emerged as major problems in the orthopedic field, and these problems were perceived to limit the lifespan of joint replacements [40]. To limit the wear particle concentration and improve the overall mechanical characteristics, efforts have been made to improve the overall characteristics of UHMWPE for hip implants. In the 90s, scientists were able to correlate changes in the physical properties of the UHMWPE with the in vivo degradation of mechanical behaviors. UHMWPE was typically sterilized by gamma irradiation, with a mean dose of 25 to 40 kGy. This process resulted in the formation of free radicals, which are the precursors of oxidation-induced embrittlement. Only in the past decade did the radiation crosslinking achieve common diffusion. This process of crosslinking combined with thermal treatment has emerged to increase wear and oxidation resistance of the polymer, and a large number of laboratory and clinical studies indicated positive outcomes [41–44]. Crosslinked polyethylene is commonly abbreviated as PEX or XLPE. Currently, there are different treatments, including irradiation and melting, irradiation and annealing, sequential irradiation with annealing, irradiation followed by mechanical deformation, and irradiation and stabilization with vitamin E [45]. Crosslinking also affects the mechanical properties of UHMWPE, corresponding usually to a decrease in the toughness, ultimate mechanical properties, stiffness, and hardness of the polymer [46]. These factors could negatively influence the device performance in vivo [47]. Free radicals may form during the manufacturing process, allowing for oxidative changes in the XLPE. As a consequence, the wear resistance of the polymer is expected to decline, the opposite behaviour constitutes a sort of paradox. Muratoglu et al. [46] studied the wear behavior of UHMWPE, finding drastic changes as a consequence of crosslinking; these authors found that this process reduces the ability of molecules to orient and reorient, inhibiting this mechanism responsible for wear. It also appeared that the level of crosslinking, found in the study, overwhelmed the effects of reduced mechanical and physical properties in controlling the wear behaviour of UHMWPE. For the best outcome, XLPE should be cross-linked at a correct level of radiation, and then re-melted to remove the free radicals [48]. The exceeding free radicals that did not react to form cross-links through irradiation must be eliminated to prevent the formation of oxidized species and their recombination. The removal can be realized through two different methods: annealing or remelting; highly cross-linked polyethylene (HXLPE) has demonstrated superior wear resistance compared to gamma-sterilized materials [46]. By annealing below the peak melting point of the polymer, some of the crystalline regions are melted and the free radical concentration is reduced, but it is still measurable. On the other hand, through post-irradiation remelting, residual free radicals are reduced to undetectable levels, as measured by state-of-the-art electron spin resonance instrument. By this process, crystallinity is reduced after the melting step due to the hindrance by the new crosslinks, so the mechanical strength and fatigue resistance of the polymer decrease [49]. Several clinical studies have been realized on the in vivo oxidation of remelted or annealed XLPEs, even if our knowledge is restricted to what might happen during the first decade of implantation [50].

Muratoglu et al. [51] analyzed retrieved XLPE acetabular liners, finding minimal oxidation, but they discovered that the oxidation increases during shelf storage in air, producing severe damage. They assumed that two mechanisms could alter the oxidative stability of UHMWPE, the in vivo cyclic loading and the absorption of lipids. Lipids are able to react with oxygen and thus extract hydrogen atoms from the polyethylene chains, provoking the initiation of free radicals.

Rinitz et al. [52] investigated short- and middle-term retrievals made of remelted and annealed HXLPEs to determine whether oxidation can lead to mechanical property changes through oxidative chain scissions.

Their studies proved crosslink density decreases, corresponding to augmented oxidation for some highly cross-linked, thermally stabilized materials. Other clinical studies highlighted fast in vivo oxidation rates of post-irradiation thermally treated retrievals [53].

Successful outcomes are reached by HXLPE liners associated with a delta ceramic femoral head, as found by Kim et al. [54], finding an annual penetration rate of the femoral head of around 0.022 mm/year. Hamai et al. compared the clinical wear rates of annealed and remelted HXLPE liners by means of radiographs on 36 matched pairs of hip explants. They found significantly greater creep in the remelted than the annealed, but no significant differences between the steady state wear rates. The retrospective study of Takada et al. [55] compared the wear behavior between the second-generation annealed and first-generation remelted HXLPEs. Involving 123 primary THA, their study confirmed excellent wear resistance of both types of HXLPE, but found that second-generation annealed HXLPE had a better wear resistance than first-generation remelted HXLPE in a short-term follow-up. Also, D'antonio et al. [56] reported the wear rate of second-generation annealed HXLPE, which compared to a conventional polyethylene, represented a reduction of 72–86% (depending on other studies results). They further found a reduction of 58%, when comparing the linear wear of the second- and first-generation annealing HXLPE.

Crystallinity of the polymer is a function of the irradiation dose and of the thermal treatment [57]. Irradiation leads to smaller chains with augmented mobility, whereas the change in crystallinity after the thermal procedure depends on the temperature reached. If the treatment is realized below the melting point of 137 °C, the chain mobility rises, yielding higher crystallinity [58,59]. If the procedure is performed at higher temperature, the crystallization of the polymer, during the cool-down to ambient temperature, occurs in the presence of cross-linking, which decreases the crystallinity of the polymer and improve the wear resistance with small changes in toughness [58].

Basically, the mechanisms by which UHMWPE improves its chains occurs via plastic deformation of the polymer, with molecular alignment in the direction of motion that results in the formation of fine, drawn-out fibrils oriented parallel to each other [60]. As a result of this arrangement, the UHMWPE wear surface may strengthen along the direction of sliding, while it weakens in the transverse direction. In light of this, there is a will to realize reinforced polymers with high strength such as self-reinforced UHMWPE [61]. This composite is basically a non-oriented matrix of UHMWPE where reinforcement particles of the same material have been dispersed, resulting in a polymer with excellent biocompatibility, increased mechanical properties and the chance to be sterilized and cross-linked such as the traditional UHMWPE [61].

In Figure 3 are presented typical PE prostheses designs.

In the recent years, a different approach was developed to stabilize polyethylene. Blending vitamin E with polymers was firstly meant as a hygienically safe stabilization, Tocopherol compounds were proposed as a stabilizer for polyolefin in the 1980s [62]. In 1994, Brach del Prever et al. [63] introduced UHMWPE blended with vitamin E for a prosthetic implant. In 2007, the first vitamin E-diffused, irradiated UHMWPE hip implant was clinically introduced in the United States (Biomet Inc., Warsaw, IN, USA) [64]. The blending led to the interruption of the oxidation cycle by decreasing the reactivity of the radical species, giving origin to a third generation of polyethylenes [64–66]. If vitamin E-stabilized, irradiated UHMWPE undergoes accelerated aging at high temperatures and/or in the presence of pure oxygen, it will be oxidatively more stable than gamma-sterilized or high-dose irradiated UHMWPE [67,68]. In vitro studies supported the hypothesis that vitamin E-blending would enhance the oxidative stability of XLPEs. There are also some drawbacks in the procedure: increasing the concentration of vitamin E in the blend is not viable, the obstacle of cross-linking in the presence of vitamin E prescribes the use of a lower concentration [69]. Therefore, a balance is needed to obtain elevate cross-linking density and high oxidative stability.

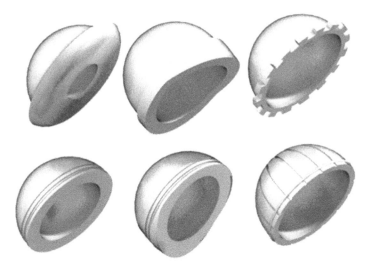

Figure 3. Some of the designs that are achieved with polyethylene for the acetabular cup.

2.3. PEEK

Polyether-ether-ketone (PEEK) is a well-known biocompatible polymer used in orthopedic applications [70]. It has been considered as an alternative joint arthroplasty bearing material due to its favorable mechanical properties and the biocompatibility of its wear debris [71]. PEEK had been used as biomaterials, in particular in the spine, since the 1980s [72,73], due to its structure that confers outstanding chemical resistance, inertness, and thermal stability for in vivo conditions. In 1998, Wang and coworkers [74] tested acetabular cups made of PEEK on a hip simulator for 10 million cycles. They observed a reduction in the wear rate of almost two orders of magnitude in comparison to a conventional UHMWPE/metal or UHMWPE/ceramic couple. However, despite the good promises deriving from in vitro, low contact stress situations, when in high contact stress environments, there are questions about the suitability of this material as acetabular cups or knee tibial components [75,76]. No clinical data of its use are available.

3. Metals

Metallic materials have wide applications in the medical and bioengineering fields and are widespread as orthopedic implants components. The most common traditional metals used for THA are stainless steels, titanium alloys (Ti6Al4V) and—mainly—cobalt-chromium-molybdenum alloys. The latter have good corrosion resistance compared to other metals, and high toughness, high wear resistance and higher hardness (HV = 350) than other metals and polymers.

3.1. Cobalt Chromium Molybdenum Alloys

MoM articulation is typically produced from cobalt-chromium-molybdenum (CoCrMo) alloys. CoCrMo alloys are composed of 58.9–69.5% Co, 27.0–30% Cr, 5.0–7.0% Mo, and small amount of other elements (Mn, Si, Ni, Fe and C). These metallic alloys can be divided in 2 categories: high-carbon alloys (carbon content >0.20%) and low-carbon alloys (carbon content <0.08%) [77,78]. In addition, metallic alloys can be manufactured using 2 different techniques such as casting and forging; the grain size of the forged alloy is typically less than 10 μm, whereas the grain size of the cast material ranges from 30 to 1000 μm [79]. Intensive studies were done on the metallurgy for CoCrMo alloys with carbon; nevertheless, there is no complete phase diagram. This is mainly due to the complex phases existing in the system. Various carbide species, such as $M_{23}C_6$, and M_6C can take place based on the heat treatment [80]. The differences in the microstructure of the carbides, their chemical composition, and nano-hardness are related to wear performances.

Cobalt and chromium are both present in the environment and in food. They are necessary to human beings as trace elements in the body but are toxic when highly concentrated. Patients with Co-Cr metal-on-metal pairings are exposed to wear with release of cobalt and chromium into the synovial fluid. These are capable of migrating to the blood before being expelled through the urine [81,82]. There is poor knowledge on the effects of circulating Co and Cr; they may affect mainly biological and cellular functions with potential effects on the immune system, mutagenesis, and carcinogenesis. In patients with metal-on-metal hip implant, elevated levels of circulating Co and Cr ions may be generated, and there is a positive linear correlation with a lymphocytic reactivity [83,84].

3.2. Other Metal Alloys

Metallic materials have high module of elasticity, which limits stress distribution from implant to bone. Therefore, new metallic components have been developed with lower elastic modulus and higher corrosion and wear resistance. There is continuous research for new metallic alloys for application in hip prostheses to obtain a better biocompatibility along with superior mechanical properties. Still, it is mandatory to find a compromise between the many optimal characteristics desired for an implant material. Co-Cr-Mo alloys have low chemical inertness but high wear resistance, whereas stainless steel alloys have low strength and ductility. Zirconium (Zr) and tantalum (Ta) are refractory metals—due to their great chemical stability and elevate melting point—and are very resistant to corrosion, due to the stability of the oxide layer. As vanadium is a relatively toxic metal, some attempts were made to replace it in the widespread Ti-6Al-4V alloys. To improve biocompatibility and mechanical resistance, this Ti-6Al-4V alloys was replaced with iron (Fe) or niobium (Nb), realizing the improved alloys Ti-5Al-2.5Fe and Ti-6Al-7Nb. These alloys with respect to the traditional Ti-6Al-4V have greater dynamic hardness and lower elastic module, allowing a better implant/bone stress distribution. A new class of titanium alloys introduced into the orthopedic field uses molybdenum in concentration greater than 10%. Its presence stabilizes the β-phase at room temperature; these are referred to as β-Ti alloys. Having 20% less elastic modulus, they behave closer to real bones and have better shaping possibilities. Femoral stems made of a β titanium alloy have been used as part of modular hip replacements since the early 2000's but were recalled in 2011 by the US Food & Drug Administration (FDA) due to elevated levels of wear debris. Yang and Hutchinson [85] found that the dry wear behaviour of a β titanium alloy (TMZF (Ti-12Mo-6Zr-2Fe (wt.%)) is very similar to that of Ti64, whereas their behaviour is completely different in simulated body fluid, where the wear of TMZF is significantly accelerated. Another recently introduced metal material is the oxidized zirconium (Oxinium, by Smith & Nephew), with a metal core and abrasion-resistant ceramic surface. The niobium alloy of zirconium has proven to decrease the UHMWPE wear rate and particle production considerably [86]. In Figure 4 it is possible to see the design of metal implants with different material renderings.

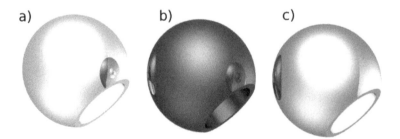

Figure 4. Metal femoral heads: (**a**) stainless-steel; (**b**) Oxinium; (**c**) CoCrMo.

The revision rate of large head metal-on-metal and resurfacing hips is significantly higher than that of conventional total hip replacements. The revision of these bearings has been linked to high wear as a consequence of edge loading, which happens when the head-cup contact patch extends over the cup rim [87]. Underwood et al. [88] highlighted that using hip implants with low clearance, having more conformal contact and so a larger contact patch, increases the risk of edge loading and therefore intense wear.

4. Ceramics

The word ceramics derives from Greek, *keramos*, meaning potter or pottery. Ceramics were defined by Kingery [89] as "the art and science of making and using solid articles, which have, as their essential component, and are composed in large part of, inorganic nonmetallic materials". It is likely to say that a ceramic is whatever material is neither a metal, a semiconductor or a polymer. Ceramics are used to build engineering components when wear resistance, hardness, strength and heat resistance are required. Ceramics were also defined as "the materials of the future", as they are derived from sand that is about 25% of the earth's crust as compared to 1% for all metals [90]. In the lasts decades, ceramic materials have exhibited great appealing and diffusion thanks to their chemical and physical characteristics, attracting the interest of biomedical scientists and companies [91]. Ceramic materials were introduced in the THA more than twenty years ago to overcome the major issue of polyethylene wear [92].

4.1. Alumina

Alumina was introduced in THA implants in 1971, when Boutin realized alumina-on-alumina hip coupling, leading to good clinical results [93,94]. Alumina ceramic has been one of the main ceramics to be used in THA, relying on its good tribological properties, meaning a favorable frictional behavior and a high wear resistance [95]. On the other hand, it has weaker mechanical resistance than other materials. It showed good performances in compression, but weak resistance to tensile stresses [96]. Alumina ceramics have been used in clinical applications for their tribological properties due to their hardness [97]. Among the ceramics, alumina is probably the most commonly used material.

The alumina used for hip replacements was different from the first generation of the material used for industrial applications. In particular, the first generation of alumina showed poor microstructure with low density, scarce purity, and large grain size. This generation of alumina was unsuited for biomedical use. The continuous efforts performed in this field allowed researchers to purify and improve this process, leading to an alumina for medical use, commercially known as Biolox® [21,92]. The ISO 6474 standard, introduced in 1980, aimed to improve the quality of alumina for THA and to decrease the fracture occurrence. Alumina performance is related to different aspects, such as the density, the purity and the grain size. The last one, in particular, influences the wear rate, as it decreases with smaller grain size [92]. In the 90's alumina hip implants were improved with the arrival of Biolox® forte on the market, which could rely on innovations in the production process to furnish much better mechanical characteristics [21,92]. It was realized using improved raw material, with smaller gain size, low level of impurities and sintered in air. Biolox® forte has a density of 3.98 g/cm^3 and grain size of 3.2 m, whereas for Biolox®, these values are 3.96 g/cm^3 and 4.2 m [98].

Recently, concerns have been raised because of some clinical reports on the presence of audible noise in some ceramic-on-ceramic THA patients [99]. The so-called "clicks" or "grinds" have been described after THA, regardless of whether metal-on-polyethylene, metal-on-metal, or ceramic-on-ceramic bearings were used [100]. The "squeak" appears to be limited, however, to hard-bearing couples. It is probably related to implant design or cup orientation and the exact etiology of squeaking is the object of debates; there is neither a specific definition for post-surgery squeaking nor a universal categorization for the sound [101].

4.2. Zirconia

Zirconia has high toughness and good mechanical properties; among all the monolithic ceramics, it has outstanding crack resistance [102]; these are the main reasons that made zirconia a very widespread alternative to alumina for THA. Firsts attempts were focused on magnesia partially stabilized zirconia (MgPSZ), that did not satisfy the wear resistance requirements [103]. Therefore, further developments were focused on yttria stabilizing oxide (Y-TZP), a ceramic that is completely formed by submicron-sized grains, representing the current standard for clinical application [104]. A picture of such a ceramic femoral head is shown in Figure 5.

Figure 5. Zirconia femoral head.

Y-TZP is composed of tetragonal grains sized less than 0.5 μm, the faction of which retained at room temperature depends on the size, the distribution and the concentration of the yttria stabilizing oxide [96]. Such microstructural parameters define the mechanical properties of the Y-TZP. The tetragonal grains can transform into monoclinic grains, producing 3–4% volume expansion [105], which is the reason behind the toughness of the ceramic and its ability to dissipate the fracture energy. When a pressure acts on grains, e.g. a crack advancing in the material, they shift to the monoclinic phase, dissipating the crack energy in two ways: the T-M transformation and the volume expansion [106]. There are also metastable tetragonal phase particles, of which formation depends on grain size, stabilizing oxide concentration and matrix constraint. Above 100°C, the metastable particles in a wet environment can spontaneously transform into monoclinic particles [107]. As the transformation progresses, a decrease in material density and in strength and toughness of the ceramic can be observed. The structure of Y-TZP at room temperature is realized by submicron sized grains that grow during the sintering; it is therefore necessary to start from submicron size powders (e.g., 0.02 μm) and to introduce some sintering aid to limit the phenomenon [9].

With respect to metals, Y-TZP shows superior wettability properties that allows for fluid film formation between the articulating surfaces of an implant. Even if in clinical practice the Y-TZP femoral heads were only coupled with UHMWPE cups, tests performed on Y-TZP vs. alumina returned positive results [108]. From the wide investigation campaign on the wear performance of UHMWPE vs. zirconia, there is a general agreement on the fact that the wear is not higher than

UHMWPE vs. alumina [109–111]. Discrepancies in results derive mostly from the differences in the bulk materials used in laboratories, in their finishes, testing procedures etc. There is great concern in the orthopedic community regarding the future of Zirconia as prosthesis. The market has decreased more than 90% between 2001 and 2002 (corresponding with the recall and abandon of Prozyr®, by Saint Gobain) [112]. More than 600000 femoral heads used in Y-TZP have been implanted worldwide, mostly in EU and US. The debate on the Y-TZP future is due to its pros and cons; it exhibits the best mechanical properties (resistance to crack propagation) but is prone to aging in the presence of water.

Zirconia manufacturers tried to shrink this problem, claiming that it was limited under in vivo conditions until 2001 when around 400 femoral heads failed in a short period. This event was related to accelerated ageing affecting two batches of Prozyr® [112]. Even if the reason was identified to be processed controlled, this event led to catastrophic impact on the use of the Y-TZP, pushing some surgeons to go back to other solutions. The ageing problem and the Prozyr® event are still an issue, and further efforts are required to gain confidence from the orthopedic society. In this way, the future seems to be based on the combination of zirconia and alumina to obtain advanced composites.

4.3. Zirconia Toughened Alumina

In the second half of the 1970s, a new class of ceramic-based composite materials developed. This new composite material was realized by introducing up to 25% wt. of zirconia into an alumina matrix; this composite material is known as zirconia toughened alumina (ZTA). The addition of a fraction of zirconia to alumina results in a composite material of increased toughness [109,110,113]. In the 2000s, the first ZTA material introduced in a clinic was a composite known under the trade name of Biolox® Delta [114]. A picture of such a ceramic femoral head is shown in Figure 6.

Figure 6. Biolox® Delta femoral head.

This material provides elevate resistance to the onset of cracking and to crack propagation [115,116]. This ZTA composite combines the best characteristics of both alumina and zirconia: the strength and toughness of alumina and the excellent wear resistance, chemical and hydrothermal stability of the alumina. This combination is realized through the uniform distribution of nano-sized particles of yttria-stabilized tetragonal zirconia (Y-TZP) in the alumina matrix. A small percentage of chromium oxide (Cr_2O_3) is added to counterbalance the hardness reduction caused by the zirconia presence. Strontium oxide (SrO) is added to the material, during the sintering process, to form strontium aluminate ($SrAl_{12-x}Cr_xO_{19}$) platelets [117]. These flat and elongated crystals dissipate cracks energy and limit their advance, as it would require extra energy for the crack to overtake the crystal. The final composite is a mixture of roughly 75% alumina, 25% zirconia, and less than 1% chromium oxide and strontium oxide [96]. Deville et al. [118] found that Alumina Y-TZP composites exhibit significant ageing, but this process was far slower than usually observed in Y-TZP ceramics, which is ascribable to the presence of the alumina. On the other side, the presence of zirconia aggregates was recognized as the main cause of ageing sensitivity [119]. Realizing an optimal dispersion at acid pH can avoid the formation of zirconia aggregates, but as soon as the percolation threshold level (16 vol.%) is exceeded, ageing cannot be avoided.

These composites achieve a fracture toughness (K_{IC}) up to 12 MPa·m$^{1/2}$ and a bending strength up to 700 MPa. Due to the different elastic moduli of the two components, cracks will tend to move across the less stiff zirconia particles, inducing their T-M phase transformation that dissipates the crack energy.

5. Wear Behavior

Among the bearing surfaces involved in total hip arthroplasty, the biomaterials are submitted to sliding friction, producing particle debris, which, in turn, initiate an inflammatory reaction ultimately leading to osteolysis [120]. Wear is defined as a cumulative surface damage phenomenon in which material is removed from a body in the form of small particles, primarily by mechanical processes [121]. The wear mechanism is the transfer of energy with removal or displacement of material and in that follows an explanation of the mechanisms of wear observed with different biomaterials.

Pertinent literature was obtained from the Scopus database. The key words "hip joint replacement," "hip prostheses," "in vitro wear," "in vivo wear," and "THA" were searched in various combinations, and results were narrowed based on relevance to this review. Only articles from peer-reviewed journals were included.

5.1. Wear of Polyethylene

The primary mechanism of wear of polyethylene in THA is adhesive/abrasive, leading to the formation of sub-micron sized particles [33]. Elongated fibrils found in retrieved acetabular elements are precursors for this wear mechanism [58]. There is proof that the morphology of UHMWPE changes due to mechanical input. For example, it has been found that the mechanical properties of the polymer are dependent on both its crystalline and amorphous phases wear is led, at a micro-scale, by cyclic plastic deformation of the articulating surface [38]. Microstructural changes are correlated with plastic deformation in UHMWPE, in that lamellar alignment has been found during tests of cyclic tension, as well as decreased crystallinity in monotonic tension and compression specimens taken past yield [122].

There are different factors that influence the UHMWPE wear; some of them are related to the material itself, other are mostly due to the whole implant design. In the first category, is the nature or quality of the powder, as well as the tensile-rupture energy, the manufacturing process and the sterilization procedure. UHMWPE components can be obtained from ram-extruded bars; this process leads to internal inconsistencies or "dead zones". The dead zones can lower the molecular weight and increase the wear rate of the final component [123]. Furthermore, the so-obtained elements tend to have micro-shred on their surface that can cause the third-body wear process. If the component is realized through heat stamping, as the melted outer layer cools, crystallization begins. The differential cooling leads to internal stresses resulting in a final element with anisotropic strength properties, vulnerable to oxidation degradation.

In the adhesion/abrasion wear mechanism, the surface conditions of the femoral head component, in particular its roughness and hardness, are key aspects. The hardness of the head material should be higher than that of the acrylic bone cement. If so, in a cemented arthroplasty, there will be less likely for third-body wear at bearing surfaces. To minimize the UHMWPE wear rate, the counter-body should be very hard and have a low contact angle (less than 70°); further, the head should be as smooth as possible and inert to oxidation.

5.2. Wear of Metals

The dynamic loading these implants undergo, together with the corrosiveness of physiological fluids can enhance the degradation processes. The combined effect of wear and corrosion does not consist of a simple sum of the two but more as a synergy realized between them called tribo-corrosion. Tribo-corrosion is defined as an "irreversible transformation of material in tribological contact caused by simultaneous physicochemical and mechanical surface interactions" [124]. In the last decades, a scaring occurrence of inflammatory reactions has been seen in patients with large head MoM THA, often with signs of tribo-corrosion at the head-neck interface. Tribo-corrosion arises not only at MoM bearing surfaces, but also at metal/metal modular junctions where micro-motions between the two components are possible.

More frequently, the wear of metal bearings can be distinguished in three main processes and their combinations: abrasive wear, due to either two or three bodies, adhesive wear and fatigue wear. However, other types of wear such as corrosive can occur. The corrosion resistance of metals relies on the passive layer formed on their surface in contact with a corrosive environment. Metals react with an oxygen-rich biological environment, realizing a thin protective oxidative coating – generally 2–5 nm thick – that limits corrosion. The oxidative layer forms immediately when exposed to in vivo conditions, but it does not last forever. Regarding the passive metals, wear can break the oxide layer on the surface, accelerating the dissolution of the base metal. The coatings can be scratched or rubbed off when surface contact happens. Even though the oxide layer spontaneously reforms, in restoring the protection of the surface, there is a rise in corrosion currents during the process, which causes the degradation of the material along with the release of metallic ions [125]. Once the film is worn out, the implant can release metal ions and particulates. The presence of these elements realizes third body wear that intensely increases wear rates. This damaging process applied on the coating, and metal ions released, and reformation of new coatings is known as *oxidative wear* [122]. The propensity of the layer to breakdown derives from the difference between the resting and breakdown potential. Regarding the CoCr alloys, the difference is high but corrosion can still happen under certain conditions. However, localized corrosion is not so common in CoCr alloys, which typically fail by trans-passive dissolution [125].

Galvanic corrosion can arise when different metals are in contact with each other, but also when the contact is between the same metal being partly under corrosion and partly under tribo-corrosion conditions. This type of galvanic contact is typical of modular implants, as in the neck-head contact.

Wear particles occurring in MoP implants are within the size range required for phagocytosis by macrophages, which is considered to be a cause of aseptic loosening [126]. On the other hand, particles generated by MoM implants belong to the nanometer scale, which reduces macrophage reaction. Nevertheless, the distribution of these particles within the body can have different biological effects and could be responsible for cytotoxicity, hypersensitivity and eventually carcinogenesis.

Investigations on retrieved 1st and 2nd generation MoM hip prostheses have shown linear penetrations of roughly 5 mm/year, which corresponds to a wear volume of approximately 1 mm^3/year, two orders of magnitude lower than conventional polyethylene acetabular cups. The wear of hard-on-hard articulations such as MoM hip prostheses has two separate stages. Elevated bedding in the wear period occurs during the first million cycles or first year in vivo. Afterwards, a lower steady-state wear period occurs as the bearing surfaces have been subjected to the self-polishing action of the metal wear particles, which may act as a solid-phase lubricant. In vitro investigations, realized by hip simulators, generally show steady-state wear rates to be lower than those reported in vivo. The wear of tested MoM hip prostheses, 1 mm^3/million cycles, is much lower than the more widespread polyethylene-on-metal bearings, 30-100 mm^3/million cycles [19].

Each type of Co-Cr alloy has different characteristics that influence the wear rates of an implant. These properties comprise carbon percentage, manufacturing procedure and surface finishing. High carbon alloys have an initial wear of 0.21 mm^3/million cycles for the cast implants and 0.24 mm^3/million cycles for the wrought implants, whereas, alloys with a low carbon concentration have a significantly greater wear rate of 0.76 mm^3/million cycles. The high percent carbon alloys show superior wear resistance as compared to the low percent carbon alloys with the assumption that there was no additional variation in other parameters.

In the human hip joint, wear can be designated as reciprocating sliding wear, because the contact area is smaller than the stroke of the wear path. Furthermore, the wear paths of the back and forth section of the cycle do not lie on the same geometrical lines, which lead to sliding wear. Even though, in sliding as well as in reciprocating sliding wear, all the other wear process—adhesion, abrasion, surface fatigue and tribochemical reactions—may be present at the same time [127].

5.3. Wear of Ceramics

Ceramic-on-ceramic implants have a life expectancy longer than implants with other combinations because of their very low wear rate. This clinical result led to the success of the ceramic implants: since 1990, alumina components were implanted more than 3.5 million times, whereas zirconia elements were used more than 600k times [128]. Nevertheless, ceramic is a brittle material and fractures can happen under adverse circumstances. Fracture probability is low (0.004–0.35% for alumina heads) but does occur [129]. The main causes of head fractures are local stress concentrations that are ascribed to taper interface contamination or damage or to loosening of the head on the taper [130,131].

Affatato et al. [113] tested different ceramic configurations, i.e., pure alumina vs. alumina composite. The wear rate was lower for the pure alumina than for the alumina composites. Still, no statistically significant differences were observed between the wear behaviours of these materials at a 95% level of confidence. In different work, Affatato and co-workers [11] carried out wear tests to compare the tribo-behaviour of different sizes of ceramic components. Two different batches of alumina Biolox® Forte (28 mm vs. 36 mm) were tested on a hip simulator under bovine calf serum for five million cycles. They found that the 36 mm Biolox® forte size showed less weight loss than the 28 mm Biolox® Forte size.

Nevelos et al. [132] studied the behavior of CoC bearings realized with hot isostatically pressed alumina and compared with the standard alumina ones. They found a reduction of the wear rate for the hot-pressed prosthesis when working under standard conditions. Different behavior was observed under Gelofusione® (4% w/v solution of succinylated gelatin) and water lubricants, where the non-hot-pressed ceramic showed a lower wear rate. Even so, the results were significantly affected by uncertainties as testified by the large error bars. It is worth noting that the wear rates reported by the authors, under standard testing conditions, were an order of magnitude lower than the majority of reported clinical wear rates for in vivo ceramic prostheses [133,134].

A summary of the in vitro tests realized on the different combinations of materials is presented in Tables 1 and 2, for soft and hard bearings, respectively.

Table 1. Soft bearings' wear rates found in vitro through simulators.

Soft Bearings	Paired Materials *	Overall Wear Rate (mm^3/Mc)	Ref.
MoP	CoCr—XLPE	6.71 ± 1.03	[135]
	Biolox®Delta—XLPE	2.0 ± 0.3 **	[136]
CoP	CoCrMo—XLPE	4.09 ± 0.64	[137]
	Alumina—XLPE	3.35 ± 0.29	[138]
	Alumina—PE	34	[139]
	ZTA—PE	80	[140]

* all the abbreviations are reported at the end of the manuscript. ** only in this case the unit of measure is mg/Mc.

Table 2. Hard bearings wear rates found *in vitro* through simulators.

Hard Bearings	Paired Materials *	Overall Wear Rate (mm^3/Mc)	Ref.
CoM	CoMplete	0.129 ± 0.096	[141]
	Biolox®Delta - CoCrMo	0.02 ± 0.01	[142]
	Biolox®Delta-CoCrMo	0.87	[28]
CoC	Biolox®Forte-Biolox®Forte	0.052	[28]
	Alumina-Alumina	0.03	[143]
	ATZ-ATZ	0.024	[143]
	ATZ-ATZ	0.06 ± 0.004	[144]
	ZTA-ZTA	0.14 ± 0.10	[144]
	ATZ-ZTA	0.18	[145]
	ATZ-Alumina	0.20	[145]
	Alumina-Alumina	0.74 ± 1.73	[144]
	Biolox®Delta-Biolox®Delta	0.10	[146]
MoM	CoCrMo-CoCrMo	0.60 ± 0.18	[142]
	CoCrMo-CoCrMo	0.11 ± 0.055	[147]

* all the abbreviations are reported at the end of the manuscript.

6. Discussion

Since its first application, THA has evolved in both terms of material and design. After a first experimental phase, that went along many failures, the UHMWPE was established as the most widespread material to be used as acetabular component. The arrival of CoCrMo destabilized its supremacy for a while but the combination of the two resulted in great pairing. Ceramics are the most recent materials introduced in the orthopaedic field, having the best tribological behavior, they rapidly achieved great success. During the 1970s and 1980s, the great majority of hip prostheses in clinical use incorporated a polyethylene acetabular liner bearing against a femoral ball of metal or ceramic. The willing to resolve the issues of hip implants pushed many researchers to study the various combinations of materials and to introduce some variation of their characteristics. These alternatives included highly cross-linked, thermally stabilized polyethylenes against metal, composite ceramics. The latter composites realized with ceramic matrix are the most successful ones.

The biomaterials used in the orthopedic field play a vital role, and their validation through in vitro tests is of paramount importance. The main objective in the field of biomaterials for hip implants is the reduction of failure incidences. We believe that knowledge of wear rate is an important aspect in the pre-clinical validation of prostheses. Wear tests are executed on materials and designs used in prosthetic hip implants to control their final quality and obtain auxiliary knowledge on the tribological processes. Researchers should not forget that other issues still impact the life expectancy of the prostheses, such as the sensitivity of the cup position and edge loading in ceramic bearings. Therefore, several steps forward are required to improve the overall performance of the implants, such as the ability to sustain high demand activities—for young patients—and preserve the bone from retro-acetabular loss.

New implant concepts, such as hip resurfacing and shorter cementless hip stems, are today mostly used in Europe and may also influence the future of hip arthroplasty. Considering that the number of patients who undergo total joint arthroplasty, and consequently revision, is increasing due to an aging population, patients remain the principal players in this process. There is also an increase in the economic health expense, so it is necessary to reduce the number of revisions to reduce these costs. Knowledge of the behavior of individual prostheses in certain clinical conditions may help in this matter. Nowadays there are many prosthetic models on the market and few scientific evidence of good methodological quality to support the use of most of them. Under these conditions, it is difficult to monitor the use of prosthetic devices and ensure the traceability of the patients in the case of adverse events.

Many countries are adopting a registry for post-marketing surveillance in order to collect data on joint prosthetic performance. Registries can be compiled at the international, national or regional level but also locally, such as in hospitals [148]. Through the registers, it is possible to evaluate the effectiveness of an implant, its lifetime and performance for the treatment of specific cases. Registries are an important tool for research; they allow the identification of patients with a certain condition or outcome for prospective observational studies of large size. In this way, the registry can educate the surgeon to select the best type of prosthesis and surgical technique. Consequently, the healthcare resource will be properly used.

7. Conclusions and Future Prospects

The future of total hip replacement should be perceived as a divergent tendency for developed and developing countries. Advances in technology, improved materials and better understanding of natural tissue reactions will certainly result in breakthroughs of implant selection. Due the ageing of the population, the number of joint replacement surgery has increased in the last years [149]. Consequently, also the number of revision surgeries is growing, as the life expectancy of patients is longer than that of prostheses [150,151].

Current trends in prosthesis design emphasise the use of biocompatible materials that are strong enough to withstand the more active lifestyles of many patients, whilst generating minimal wear debris. As the main issue affecting the long term durability of prosthesis is wear and the propagation of wear particles, vast research is currently being undertaken to improve such biomaterials to give an "infinitive prosthesis life". Analysis of component wear is therefore essential for future progress; retrieval analysis of a well-functioning bearing prosthesis could help in improved the biomaterials. Controversy regarding the safety of metal-on-metal bearing surfaces still remains, particularly in relation to metal ion release and potential hypersensitivity reactions [152–154]. Ceramic-ceramic implants have been demonstrated to provide the lowest wear rates in comparison to other material options possible for ceramic-on-ceramic THA [9,98,155]. Trends in material development are also strongly influenced by the desire to improve hip function and stability through the use of increased head diameters [97]. Today, there is a large number of prosthetic models on the market and limited scientific evidence of good methodological quality to support their usage; the expected costs of treatment in a decade perspective amount to a fraction of what they turned out to be. Worldwide, countries should develop strategies to tackle the problem of increasing demand for medical services in a more simplified and inexpensive way, as they may not even be capable of absorbing the technology in the absence of infrastructure, lack of training and know-how. Prevention, i.e., appropriate dietary and lifestyle modifications, may be important to reduce hip implants. In addition, as mentioned above, countries should adopt registries for post-marketing surveillance. Such registries should collect all data on joint prostheses performance in order to evaluate the effectiveness of an implant, its lifetime and performance for the treatment of specific cases. In this way, the registry can educate the surgeon on the best type of prosthesis and surgical technique or to improve preoperative planning [3]. Consequently, the healthcare resource will be properly used. In conclusion, based on the increase in hip implants in young and older patients, the development of new biomaterials correlated with the lower wear-rate, and the systematic collection of limited essential information on the surgery and the definition of a single endpoint, the failure of the system and its replacement, allow us to monitor the device over time after its market introduction. This may help the surgeons to improve the quality life of the patient in the near future.

Funding: This research was partially funded by the Italian Programme of Donation for Research "5 per mille", anno 2016 - redditi 2015, n. Cardinis 7160.

Acknowledgments: The authors would like to thank Luigi Lena (IRCCS – Istituto Ortopedico Rizzoli, Bologna-Italy) for his help with the original pictures.

Conflicts of Interest: The authors declare no conflict of interest.

Abbreviations

Alumina toughened	ATZ
Ceramic-on-ceramic	CoC
Ceramic-on-metal from	CoMplete
Cross-linked	XLPE
Metal-on-metal	MoM
Metal-on-polyethylene	MoP
Polytetrafluoroethylene	PTFE
Polyetheretherketone	PEEK
Total hip arthroplasty	THA
Ultra-high molecular weight polyethylene	UHMWPE
Zirconia toughened alumina	ZTA

References

1. Learmonth, I.D.; Young, C.; Rorabeck, C. The operation of the century: total hip replacement. *Lancet* **2007**, *370*, 1508–1519. [CrossRef]
2. Aherwar, A.; Singh, A.K.; Patnaik, A. Current and future biocompatibility aspects of biomaterials for hip prosthesis. *AIMS Bioeng.* **2015**, *3*, 23–43. [CrossRef]
3. Affatato, S. *Perspectives in Total Hip Arthroplasty: Advances in Biomaterials and Their Tribological Interactions*; Affatato, S., Ed.; Elsevier Science: Amsterdam, The Netherlands, 2014; ISBN 1782420398.
4. Gomez, P.; Morcuende, J.A. Early attempts at hip arthroplasty-1700s to 1950s. *Iowa Orthop. J.* **2005**, *25*, 25–29. [PubMed]
5. Pramanik, S.; Agarwal, A.K.; Rai, K.N. Chronology of Total Hip Joint Replacement and Materials Development. *Trends Biomater. Artif. Organs* **2005**, *19*, 15–26.
6. Knight, S.R.; Aujla, R.; Biswas, S.P. Total Hip Arthroplasty - over 100 years of operative history. *Orthop. Rev. (Pavia)* **2011**, *3*. [CrossRef] [PubMed]
7. Muster, D. Themistocles Gluck, Berlin 1890: A pioneer of multidisciplinary applied research into biomaterials for endoprostheses. *Bull. Hist. Dent.* **1990**, *38*, 3–6. [PubMed]
8. Hernigou, P. Smith-Petersen and early development of hip arthroplasty. *Int. Orthop.* **2014**, *38*, 193–198. [CrossRef] [PubMed]
9. Zivic, F.; Affatato, S.; Trajanovic, M.; Schnabelrauch, M.; Grujovic, N. *Biomaterials in Clinical Practice: Advances in Clinical Research and Medical Devices*; Springer: Berlin, Germany, 2018; ISBN 3319680250.
10. McKee, G.K. Total hip replacement - past, present and future. *Biomaterials* **1982**, *3*, 130–135. [CrossRef]
11. Affatato, S.; Spinelli, M.; Squarzoni, S.; Traina, F.; Toni, A. Mixing and matching in ceramic-on-metal hip arthroplasty: an in-vitro hip simulator study. *J. Biomech.* **2009**, *42*, 2439–2446. [CrossRef]
12. Fisher, J.; Firkins, P.J.; Tipper, J.L.; Ingham, E.; Stone, M.H.; Farrar, R. In-vitro wear performance of contemporary alumina: Alumina bearing couple under anatomically-relevant hip joint simulation. In *Reliability and Long Term Results of Ceramics in Orthopedics*; Toni, A., Willmann, G., Eds.; Thieme Verlag: Stuttgart, Germany, 2001; pp. 1291–1298.
13. Triclot, P. Metal-on-metal: History, state of the art (2010). *Int. Orthop.* **2011**, *2*, 201–206. [CrossRef]
14. Kumar, N.; Arora, G.N.C.; Datta, B. Bearing surfaces in hip replacement - Evolution and likely future. *Med. J. Armed Forces India* **2014**, *70*, 371–376. [CrossRef] [PubMed]
15. Molli, R.G.; Lombardi, A.V.; Berend, K.R.; Adams, J.B.; Sneller, M.A. Metal-on-metal vs Metal-on-improved polyethylene bearings in total hip arthroplasty. *J. Arthroplast.* **2011**, *6*, 8–13. [CrossRef] [PubMed]
16. Topolovec, M.; Cör, A.; Milošev, I. Metal-on-metal vs. metal-on-polyethylene total hip arthroplasty tribological evaluation of retrieved components and periprosthetic tissue. *J. Mech. Behav. Biomed. Mater.* **2014**, *34*, 243–252. [CrossRef] [PubMed]
17. Huang, D.C.T.; Tatman, P.; Mehle, S.; Gioe, T.J. Cumulative revision rate is higher in metal-on-metal THA than metal-on-polyethylene THA: Analysis of survival in a community registry. *Clin. Orthop. Relat. Res.* **2013**, *471*, 1920–1925. [CrossRef] [PubMed]

18. National Joint Registry for England, Wales and Northern Ireland. *11th Annual Report 2014*. 2014. Available online: http://www.njrcentre.org.uk/njrcentre/News-and-Events/NJR-11th-Annual-Report (accessed on 1 February 2019).

19. Fisher, J.; Hu, X.Q.; Stewart, T.D.; Williams, S.; Tipper, J.L.; Ingham, E.; Stone, M.H.; Davies, C.; Hatto, P.; Bolton, J.; Riley, M.; Hardaker, C.; Isaac, G.H.; Berry, G. Wear of surface engineered metal-on-metal hip prostheses. *J. Mater. Sci. Mater. Med.* **2004**, *15*, 225–235. [CrossRef] [PubMed]

20. Hu, D.; Tie, K.; Yang, X.; Tan, Y.; Alaidaros, M.; Chen, L. Comparison of ceramic-on-ceramic to metal-on-polyethylene bearing surfaces in total hip arthroplasty: a meta-analysis of randomized controlled trials. *J. Orthop. Surg. Res.* **2015**, *10*, 22. [CrossRef] [PubMed]

21. Bader, R.; Willmann, G. Ceramic cups for hip endoprostheses. 6: Cup design, inclination and antetorsion angle modify range of motion and impingement. *Biomed. Tech.* **1999**, *44*, 212–219.

22. Henssge, E.J.; Bos, I.; Willman, G. Al2O3 against Al2O3 combination in hip endoprostheses. Histological investigations with semiquantitative grading of revision and autopsy cases and abrasion measures. *J. Mater. Sci. Mater. Med.* **1994**, *5*, 657–661. [CrossRef]

23. Macchi, F.; Willman, G. Allumina Biolox forte: evoluzione, stato dell'arte e affidabilità. *Lo Scalpello* **2001**, *15*, 99–106.

24. Morrison, J.C.; Ward, D.; Bierbaum, B.E.; Nairus, J.; Kuesis, D. Ceramic-on-ceramic bearings in total hip arthroplasty. *Clin. Orthop. Relat. Res.* **2002**, *405*, 158–163. [CrossRef]

25. Barnes, C.L.; DeBoer, D.; Corpe, R.S.; Nambu, S.; Carroll, M.; Timmerman, I. Wear performance of large-diameter differential-hardness hip bearings. *J. Arthroplast.* **2008**, *23*, 56–60. [CrossRef]

26. Sauvé, P.; Mountney, J.; Khan, T.; De Beer, J.; Higgins, B.; Grover, M. Metal ion levels after metal-on-metal Ring total hip replacement: a 30-year follow-up study. *J. Bone Jt. Surg. Br.* **2007**, *89*, 586–590. [CrossRef] [PubMed]

27. Toni, A.; Traina, F.; Stea, S.; Sudanese, A.; Visentin, M.; Bordini, B.; Squarzoni, S. Early diagnosis of ceramic liner fracture. Guidelines based on a twelve-year clinical experience. *J. Bone Jt. Surg. Am.* **2006**, *88* (Suppl. 4), 55–63. [CrossRef]

28. Affatato, S.; Spinelli, M.; Zavalloni, M.; Traina, F.; Carmignato, S.; Toni, A. Ceramic-on-metal for total hip replacement: mixing and matching can lead to high wear. *Artif. Organs* **2010**, *34*, 319–323. [CrossRef] [PubMed]

29. Orishimo, K.F.; Claus, A.M.; Sychterz, C.J.; Engh, C.A. Relationship between polyethylene wear and osteolysis in hips with a second-generation porous-coated cementless cup after seven years of follow-up. *J. Bone Joint Surg. Am.* **2003**, *85-A*, 1095–1099. [CrossRef] [PubMed]

30. Bozic, K.J.; Kurtz, S.M.; Lau, E.; Ong, K.; Vail, T.P.; Berry, D.J. The Epidemiology of Revision Total Hip Arthroplasty in the United States. *J. Bone Jt. Surgery-American Vol.* **2009**, *91*, 128–133. [CrossRef] [PubMed]

31. Ramakrishna, S. *Biomaterials: A Nano Approach*; CRC Press: Boca Raton, FL, USA, 2010.

32. Stauffer, R.N. Ten-year follow-up study of total hip replacement. *J. Bone Jt. Surg. Am.* **1982**, *64*, 983–990. [CrossRef]

33. Sinha, R.K. *Hip Replacement: Current Trends and Controversies*; Marcel Dekker: New York City, NY, USA, 2002.

34. Maguire, J.K.; Coscia, M.F.; Lynch, M.H. Foreign Body Reaction to Polymeric Debris Following Total Hip Arthroplasty. *Clin. Orthop. Relat. Res.* **1987**, *216*, 213–223. [CrossRef]

35. Charnley, J. Tissue reaction to the polytetrafluoroethylene. *Lancet* **1963**, *II*, 1379. [CrossRef]

36. Schreiber, A.; Huggler, A.H.; Dietschi, C.; Jacob, H. Complications After Joint Replacement — Longterm Follow-Up, Clinical Findings, and Biomechanical Research. In *Engineering in Medicine*; Springer: Berlin, Heidelberg, 1976; pp. 187–202.

37. Wroblewski, B.M.; Fleming, P.A.; Siney, P.D. *Charnley Low-Frictional Torque Arthroplasty of the Hip*; Springer: Berlin, Germany, 1999; Volume 81.

38. Sobieraj, M.C.; Rimnac, C.M. Ultra high molecular weight polyethylene: mechanics, morphology, and clinical behavior. *J. Mech. Behav. Biomed. Mater.* **2009**, *2*, 433–443. [CrossRef]

39. *ISO 11542-1:2001-Plastics—Ultra-High-Molecular-Weight Polyethylene (PE-UHMW) Moulding and Extrusion Materials—Part 1: Designation System and Basis for Specifications*; International Organization for Standardization: Geneva, Switzerland, 2001.

40. Harris, W.H. Wear and periprosthetic osteolysis: the problem. *Clin. Orthop. Rel. Res.* **2001**, *393*, 66–70. [CrossRef]

41. Wroblewski, B.M.; Siney, P.D.; Dowson, D.; Collins, S.N. Prospective clinical and joint simulator studies of a new total hip arthroplasty using alumina ceramic heads and cross-linked polyethylene cups. *J. Bone Jt. Surg. Br.* **1996**, *78*, 280–285. [CrossRef] [PubMed]

42. McKellop, H.; Shen, F.; Lu, B.; Campbell, P.; Salovey, R. Development of an extremely wear-resistant ultra high molecular weight polythylene for total hip replacements. *J. Orthop. Res.* **1999**, *17*, 157–167. [CrossRef] [PubMed]

43. Gul, R.M. *Improved UHMWPE for Use in Total Joint Replacement*; Dept. of Materials Science and Engineering, Massachusetts Institute of Technology: Cambridge, MA, USA, 1997.

44. Shen, F.-W.; McKellop, H.A.; Salovey, R. Irradiation of chemically crosslinked ultrahigh molecular weight polyethylene. *J. Polym. Sci. Part B Polym. Phys.* **1996**, *34*, 1063–1077. [CrossRef]

45. Muratoglu, O.K.; Bragdon, C.R. Highly Cross-Linked and Melted UHMWPE. In *UHMWPE Biomaterials Handbook: Ultra High Molecular Weight Polyethylene in Total Joint Replacement and Medical Devices*; Kurtz, S.M., Ed.; William Andrew: Norwich, NY, USA, 2015; ISBN 0323354351.

46. Muratoglu, O.K.; Bragdon, C.R.; O'Connor, D.O.; Jasty, M.; Harris, W.H.; Gul, R.; McGarry, F. Unified wear model for highly crosslinked ultra-high molecular weight polyethylenes (UHMWPE). *Biomaterials* **1999**, *20*, 1463–1470. [CrossRef]

47. Harris, W.H.; Muratoglu, O.K. A Review of Current Cross-linked Polyethylenes Used in Total Joint Arthroplasty. *Clin. Orthop. Relat. Res.* **2005**, *430*, 46–52. [CrossRef]

48. Burnett, S.J.; Abos, D. Total hip arthroplasty: Techniques and results. *BB Med. J.* **2010**, *52*, 455–464.

49. Oral, E.; Ghali, B.W.; Muratoglu, O.K. The elimination of free radicals in irradiated UHMWPEs with and without vitamin e stabilization by annealing under pressure. *J. Biomed. Mater. Res. Part B Appl. Biomater.* **2011**, *97 B*, 167–174. [CrossRef]

50. Puppulin, L.; Miura, Y.; Casagrande, E.; Hasegawa, M.; Marunaka, Y.; Tone, S.; Sudo, A.; Pezzotti, G. Validation of a protocol based on Raman and infrared spectroscopies to nondestructively estimate the oxidative degradation of UHMWPE used in total joint arthroplasty. *Acta Biomater.* **2016**, *38*, 168–178. [CrossRef]

51. Muratoglu, O.K.; Wannomae, K.K.; Rowell, S.L.; Micheli, B.R.; Malchau, H. Ex Vivo Stability Loss of Irradiated and Melted Ultra-High Molecular Weight Polyethylene. *JBJS* **2010**, *92*, 2809–2816. [CrossRef]

52. Reinitz, S.D.; Currier, B.H.; Levine, R.A.; Van Citters, D.W. Crosslink density, oxidation and chain scission in retrieved, highly cross-linked UHMWPE tibial bearings. *Biomaterials* **2014**, *35*, 4436–4440. [CrossRef] [PubMed]

53. Currier, B.H.; Currier, J.H.; Mayor, M.B.; Lyford, K.A.; Van Citters, D.W.; Collier, J.P. In Vivo Oxidation of γ-Barrier–Sterilized Ultra–High-Molecular-Weight Polyethylene Bearings. *J. Arthroplast.* **2007**, *22*, 721–731. [CrossRef] [PubMed]

54. Kim, Y.-H.; Park, J.-W.; Kim, J.-S. Alumina Delta-on-Highly Crosslinked-Remelted Polyethylene Bearing in Cementless Total Hip Arthroplasty in Patients Younger than 50 Years. *J. Arthroplast.* **2016**, *31*, 2800–2804. [CrossRef] [PubMed]

55. Takada, R.; Jinno, T.; Koga, D.; Miyatake, K.; Muneta, T.; Okawa, A. Comparison of wear rate and osteolysis between second-generation annealed and first-generation remelted highly cross-linked polyethylene in total hip arthroplasty. A case control study at a minimum of five years. *Orthop. Traumatol. Surg. Res.* **2017**, *103*, 537–541. [CrossRef] [PubMed]

56. D'Antonio, J.A.; Capello, W.N.; Ramakrishnan, R. Second-generation annealed highly cross-linked polyethylene exhibits low wear. *Clin. Orthop. Relat. Res.* **2012**, *470*, 1696–1704. [CrossRef] [PubMed]

57. Bhateja, S.K. Radiation-induced crystallinity changes in linear polyethylene: Influence of aging. *J. Appl. Polym. Sci.* **1983**, *28*, 861–872. [CrossRef]

58. Muratoglu, O.K.; Bragdon, C.R.; O'Connor, D.O.; Skehan, H.; Delany, J.; Jasty, M.; Harris, W.H. The Effect Of Temperature On Radiation Crosslinking Of Uhmwpe For Use In Total Hip Arthroplasty. In *46th Annual Meeting*; Orthopaedic Research Society: Orlando, FL, USA, 2000.

59. Oral, E.; Beckos, C.G.; Muratoglu, O.K. Free Radical Elimination In Irradiated Uhmwpe Through Crystal Mobility In Phase Transition To The Hexagonal Phase. *Polymer (Guildf)* **2008**, *49*, 4733–4739. [CrossRef]

60. Bracco, P.; Bellare, A.; Bistolfi, A.; Affatato, S. Ultra-High Molecular Weight Polyethylene: Influence of the Chemical, Physical and Mechanical Properties. *Materials (Basel)* **2017**, *10*, 791. [CrossRef]

61. Deng, M.; Shalaby, S.W. Properties of self-reinforced ultra-high-molecular-weight polyethylene composites. *Biomaterials* **1997**, *18*, 645–655. [CrossRef]

62. Dolezel, B.; Adamirova, L. Method of hygienically safe stabilization of polyolefines against thermoxidative and photooxidative degradation. *Czechoslovakian Social. Repub.* **1982**, *221*, 403.

63. Brach del Prever, E.M.; Camino, G.; Costa, L.; Crova, M.; Dallera, A.; Gallianro, P. Impianto Protesico Contenente un Componente di Materiale Plastico. Italian Patent 1271590, May 1994.

64. Bracco, P.; Oral, E. Vitamin E-stabilized UHMWPE for Total Joint Implants: A Review. *Clin. Orthop. Relat. Res.* **2011**, *469*, 2286–2293. [CrossRef] [PubMed]

65. Affatato, S.; De Mattia, J.S.; Bracco, P.; Pavoni, E.; Taddei, P. Wear performance of neat and vitamin E blended highly cross-linked PE under severe conditions: The combined effect of accelerated ageing and third body particles during wear test. *J. Mech. Behav. Biomed. Mater.* **2016**, *64*, 240–252. [CrossRef] [PubMed]

66. Kurtz, S.; Bracco, P.; Costa, L. Vitamin-e-Blended UHMWPE Biomaterials. In *UHMWPE Biomaterials Handbook*; Elsevier: Amsterdam, The Netherlands, 2009; pp. 237–247, ISBN 9780123747211.

67. Oral, E.; Rowell, S.L.; Muratoglu, O.K. The effect of α-tocopherol on the oxidation and free radical decay in irradiated UHMWPE. *Biomaterials* **2006**, *27*, 5580–5587. [CrossRef] [PubMed]

68. Oral, E.; Wannomae, K.K.; Hawkins, N.; Harris, W.H.W.H.; Muratoglu, O.K.O.K. α-Tocopherol-doped irradiated UHMWPE for high fatigue resistance and low wear. *Biomaterials* **2004**, *25*, 5515–5522. [CrossRef] [PubMed]

69. Kurtz, S.M.; Bracco, P.; Costa, L.; Oral, E.; Muratoglu, O.K. Vitamin E-Blended UHMWPE Biomaterilas. In *UHMWPE Biomaterials Handbook: Ultra High Molecular Weight Polyethylene in Total Joint Replacement and Medical Devices*; Kurtz, S.M., Ed.; Elsevier: Norwich, NY, USA, 2015; p. 840, ISBN 0323354351.

70. Anguiano-Sanchez, J.; Martinez-Romero, O.; Siller, H.R.; Diaz-Elizondo, J.A.; Flores-Villalba, E.; Rodriguez, C.A. Influence of PEEK Coating on Hip Implant Stress Shielding: A Finite Element Analysis. *Comput. Math. Methods Med.* **2016**, *2016*, 6183679. [CrossRef]

71. Cowie, R.M.; Briscoe, A.; Fisher, J.; Jennings, L.M. PEEK-OPTIMA ™ as an alternative to cobalt chrome in the femoral component of total knee replacement: A preliminary study. *Proc. Inst. Mech. Eng. Part H J. Eng. Med.* **2016**, *230*, 1008–1015. [CrossRef]

72. Kurtz, S.; Devine, J.N. PEEK biomaterials in trauma, orthopedic, and spinal implants. *Biomaterials* **2007**, *28*, 4845–4869. [CrossRef]

73. Kurtz, S. *PEEK Biomaterials Handbook*; Elsevier: Amsterdam, The Netherlands, 2012; ISBN 9781437744637.

74. Wang, A.; Lin, R.; Polineni, V.K.; Essner, A.; Stark, C.; Dumbleton, J.H. Carbon fiber reinforced polyether ether ketone composite as a bearing surface for total hip replacement. *Tribol. Int.* **1998**, *31*, 661–667. [CrossRef]

75. Grupp, T.M.; Utzschneider, S.; Schröder, C.; Schwiesau, J.; Fritz, B.; Maas, A.; Blömer, W.; Jansson, V. Biotribology of alternative bearing materials for unicompartmental knee arthroplasty. *Acta Biomater.* **2010**, *6*, 3601–3610. [CrossRef]

76. Brockett, C.L.; Carbone, S.; Abdelgaied, A.; Fisher, J.; Jennings, L.M. Influence of contact pressure, cross-shear and counterface material on the wear of PEEK and CFR-PEEK for orthopaedic applications. *J. Mech. Behav. Biomed. Mater.* **2016**, *63*, 10–16. [CrossRef] [PubMed]

77. Affatato, S.; Traina, F.; Ruggeri, O.; Toni, A. Wear of metal-on-metal hip bearings: Metallurgical considerations after hip simulator studies. *Int. J. Artif. Organs* **2011**, *34*, 1155–1164. [CrossRef]

78. Ihaddadene, R.; Affatato, S.; Zavalloni, M.; Bouzid, S.; Viceconti, M. Carbon composition effects on wear behaviour and wear mechanisms of metal-on-metal hip prosthesis. *Comput. Methods Biomech. Biomed. Engin.* **2011**, *14*, 33–34. [CrossRef]

79. Davis, J.R. *ASM Specialty Handbook: Nickel, Cobalt, and Their Alloys*; ASM International: Almere, The Netherlands, 2000; ISBN 978-0-87170-685-0.

80. Clemow, A.J.T.; Daniell, B.L. Solution treatment behavior of Co-Cr-Mo alloy. *J. Biomed. Mater. Res.* **1979**, *13*, 265–279. [CrossRef]

81. Delaunay, C.; Petit, I.; Learmonth, I.D.; Oger, P.; Vendittoli, P.A. Metal-on-metal bearings total hip arthroplasty: The cobalt and chromium ions release concern. *Orthop. Traumatol. Surg. Res.* **2010**, *96*, 894–904. [CrossRef]

82. Brodner, W.; Bitzan, P.; Meisinger, V.; Kaider, A.; Gottsauner-Wolf, F.; Kotz, R. Elevated serum cobalt with metal-on-metal articulating surfaces. *J. Bone Jt. Surg. Br.* **1997**, *79*, 316–321. [CrossRef] [PubMed]

83. Hallab, N.J.; Anderson, S.; Stafford, T.; Glant, T.; Jacobs, J.J. Lymphocyte responses in patients with total hip arthroplasty. *J. Orthop. Res.* **2005**, *23*, 384–391. [CrossRef] [PubMed]

84. Jacobs, J.J.; Hallab, N.J.; Skipor, A.K.; Urban, R.M. Metal degradation products: a cause for concern in metal-metal bearings? *Clin. Orthop. Relat. Res.* **2003**, *417*, 139–147. [CrossRef]

85. Yang, X.; Hutchinson, C.R. Corrosion-wear of β-Ti alloy TMZF (Ti-12Mo-6Zr-2Fe) in simulated body fluid. *Acta Biomater.* **2016**, *42*, 429–439. [CrossRef]

86. Good, V.; Ries, M.; Barrack, R.L.; Widding, K.; Hunter, G.; Heuer, D. Reduced Wear With Oxidized Zirconium Femoral Heads. *J. Bone Jt. Surgery-american Vol.* **2003**, *85*, 105–110. [CrossRef]

87. Langton, D.J.; Jameson, S.S.; Joyce, T.J.; Hallab, N.J.; Natu, S.; Nargol, A.V.F. Early failure of metal-on-metal bearings in hip resurfacing and large-diameter total hip replacement. *J. Bone Jt. Surg. Br.* **2010**, *92-B*, 38–46. [CrossRef]

88. Underwood, R.J.; Zografos, A.; Sayles, R.S.; Hart, A.; Cann, P. Edge loading in metal-on-metal hips: low clearance is a new risk factor. *Proc. Inst. Mech. Eng. H* **2012**, *226*, 217–226. [CrossRef]

89. Kingery, W.D. *Introduction to Ceramics*; John Wiley: Hoboken, NJ, USA, 2016; ISBN 047155460X.

90. Hsu, S.M.; Shen, M.C. Ceramic Wear Maps. *Wear* **1996**, *200*, 154–175. [CrossRef]

91. Cuckler, J.M.; Bearcroft, J.; Asgian, C.M. Femoral head technologies to reduce polyethylene wear in total hip arthroplasty. *Clin. Orthop. Relat. Res.* **1995**, *317*, 57–63.

92. Willman, G. Ceramics for total hip replacement - what a surgeon should know. *Orthopedics* **1998**, *21*, 173–177.

93. Boutin, P. [Alumina and its use in surgery of the hip. (Experimental study)]. *Press. Med.* **1971**, *79*, 639–640.

94. Boutin, P.; Blanquaert, D. A study of the mechanical properties of alumina-on-alumina total hip prosthesis. *Rev. Chir. Orthop. Reparatrice Appar. Mot.* **1981**, *67*, 279–287.

95. Hamadouche, M.; Sedel, L. Ceramics in orthopaedics. *J. Bone Jt. Surg. Br.* **2000**, *82*, 1095–1099. [CrossRef]

96. Piconi, C.; Maccauro, G.; Muratori, F.; Branch Del Prever, E. Alumina and zirconia ceramics in joint replacements. *JABB* **2003**, *1*, 19–32.

97. Affatato, S.; Ruggiero, A.; Merola, M. Advanced biomaterials in hip joint arthroplasty. A review on polymer and ceramics composites as alternative bearings. *Compos. Part B Eng.* **2015**, *83*, 276–283. [CrossRef]

98. Affatato, S.; Jaber, S.A.; Taddei, P. Ceramics for hip joint replacement. In *Biomaterials in Clinical Practice*; Zivic, F., Ed.; Springer: Berlin, Germany, 2017.

99. Affatato, S.; Traina, F.; Mazzega-Fabbro, C.; Sergo, V.; Viceconti, M. Is ceramic-on-ceramic squeaking phenomenon reproducible in vitro? A long-term simulator study under severe conditions. *J. Biomed Mater Res. B Appl. Biomater* **2009**, *91*, 264–271. [CrossRef]

100. Jacobs, B.C.A.; Greenwald, A.S.; Oxon, D.; Anderson, P.A.; Matthew, J.; Mihalko, W.M. *Squeaky Hips Make Media, Medical Noise*; American Academy of Orthopaedic Surgeons: Rosemont, IL, USA, 2008.

101. Wu, G.L.; Zhu, W.; Zhao, Y.; Ma, Q.; Weng, X.S. Hip squeaking after ceramic-on-ceramic total hip arthroplasty. *Chin. Med. J. (Engl.)* **2016**, *129*, 1861–1866. [CrossRef]

102. De Aza, A.H.; Chevalier, J.; Fantozzi, G.; Schehl, M.; Torrecillas, R. Crack growth resistance of alumina, zirconia and zirconia toughened alumina ceramics for joint prostheses. *Biomaterials* **2002**, *23*, 937–945. [CrossRef]

103. Rahaman, M.N.; Yao, A.; Bal, B.S.; Garino, J.P.; Ries, M.D. Ceramics for Prosthetic Hip and Knee Joint Replacement. *J. Am. Ceram. Soc.* **2007**, *90*, 1965–1988. [CrossRef]

104. Kelly, J.R.; Denry, I. Stabilized zirconia as a structural ceramic: An overview. *Dent. Mater.* **2008**, *24*, 289–298. [CrossRef]

105. VanValzah, J.R.; Eaton, H.E. Cooling rate effects on the tetragonal to monoclinic phase transformation in aged plasma-sprayed yttria partially stabilized zirconia. *Surf. Coat. Technol.* **1991**, *46*, 289–300. [CrossRef]

106. Platt, P.; Frankel, P.; Gass, M.; Howells, R.; Preuss, M. Finite element analysis of the tetragonal to monoclinic phase transformation during oxidation of zirconium alloys. *J. Nucl. Mater.* **2014**, *454*, 290–297. [CrossRef]

107. Maccauro, G.; Rossi Iommetti, P.; Raffaelli, L.; Manicone, P.F. Alumina and Zirconia Ceramic for Orthopaedic and Dental Devices. In *Biomaterials Applications for Nanomedicine*; Pignatello, R., Ed.; InTech: London, UK, 2011; Volume 3, p. 485, ISBN 9533076615.

108. Pitto, R.P.; Blanquaert, D.; Hohmann, D. Alternative bearing surfaces in total hip arthroplasty: Zirconia-alumina pairing. Contribution or caveat? *Acta Orthop. Belg.* **2002**, *68*, 242–250.

109. Affatato, S.; Testoni, M.; Cacciari, G.L.; Toni, A. Mixed oxides prosthetic ceramic ball heads. Part 2: effect of the ZrO_2 fraction on the wear of ceramic on ceramic joints. *Biomaterials* **1999**, *20*, 971–975. [CrossRef]

110. Affatato, S.; Goldoni, M.; Testoni, M.; Toni, A. Mixed oxides prosthetic ceramic ball heads. Part 3: Effect of the ZrO2 fraction on the wear of ceramic on ceramic hip joint prostheses. A long-term in vitro wear study. *Biomaterials* **2001**, *22*, 717–723. [CrossRef]

111. Piconi, C.; Burger, W.; Richter, H.G.; Cittadini, A.; Maccauro, G.; Covacci, V.; Bruzzese, N.; Ricci, G.A.; Marmo, E. Y-TZP ceramics for artificial joint replacements. *Biomaterials* **1998**, *19*, 1489–1494. [CrossRef]

112. Chevalier, J. What future for zirconia as a biomaterial? *Biomaterials* **2006**, *27*, 535–543. [CrossRef]

113. Affatato, S.; Torrecillas, R.; Taddei, P.; Rocchi, M.; Fagnano, C.; Ciapetti, G.; Toni, A. Advanced nanocomposite materials for orthopaedic applications. I. A long-term in vitro wear study of zirconia-toughened alumina. *J. Biomed. Mater. Res. Part B Appl. Biomater.* **2006**, *78*, 76–82. [CrossRef]

114. Gadow, R.; Kern, F. Novel zirconia-alumina nanocomposites combining high strength and toughness. In *Advanced Engineering Materials*; John Wiley & Sons: Hoboken, NJ, USA, 2010; Volume 12, pp. 1220–1223.

115. Dickinson, A.; Browne, M.; Jeffers, J.; Taylor, A. Development of a Ceramic Acetabular Cup Design for Improved in vivo Stability and Integrity. In *BIOLOX Symposium*; Sprimger: Edinburgh, UK, 2009; Volume 13, p. 227.

116. Merkert, P. Next generation ceramic bearings. In *Bioceramics in Joint Arthroplasty*; Springer: Berlin, Germany, 2003; pp. 123–125.

117. Affatato, S.; Modena, E.; Toni, A.; Taddei, P. Retrieval analysis of three generations of Biolox®femoral heads: Spectroscopic and SEM characterisation. *J. Mech. Behav. Biomed. Mater.* **2012**, *13*, 118–128. [CrossRef]

118. Deville, S.; Chevalier, J.; Fantozzi, G.; Bartolomé, J.F.; Requena, J.; Moya, J.S.; Torrecillas, R.; Díaz, L.A. Low-temperature ageing of zirconia-toughened alumina ceramics and its implication in biomedical implants. *J. Eur. Ceram. Soc.* **2003**, *23*, 2975–2982. [CrossRef]

119. Gutknecht, D.; Chevalier, J.; Garnier, V.; Fantozzi, G. Key role of processing to avoid low temperature ageing in alumina zirconia composites for orthopaedic application. *J. Eur. Ceram. Soc.* **2007**, *27*, 1547–1552. [CrossRef]

120. Massin, P.; Achour, S. Wear products of total hip arthroplasty: The case of polyethylene Produits d ' usure des arthroplasties totales de hanche: le cas du polyéthylène. *Morphologie* **2017**, *101*, 1–8. [CrossRef]

121. Guy, R.; Nockolds, C.; Phillips, M.; Roques-Carmes, C. Implications of Polishing Techniques in Quantitative X-Ray Microanalysis. *J. Res. Natl. Inst. Stand. Technol.* **2002**, *107*, 639–662.

122. Butler, M.F.; Donald, A.M.; Ryan, A.J. Time resolved simultaneous small- and wide-angle x-ray scattering during polyethylene deformation-II. Cold drawing of linear polyethylene. *Polymer1* **1998**, *39*, 781–792. [CrossRef]

123. Lewis, G. Polyethylene wear in total hip and knee arthroplasties. *J. Biomed. Mater. Res.* **1997**, *38*, 55–75. [CrossRef]

124. Landolt, D.; Mischler, S.; Stemp, M. Electrochemical methods in tribocorrosion: A critical appraisal. *Electrochim. Acta* **2001**, *46*, 3913–3929. [CrossRef]

125. Hodgson, A.W.E.; Kurz, S.; Virtanen, S.; Fervel, V.; Olsson, C.-O.A.; Mischler, S. Passive and transpassive behaviour of CoCrMo in simulated biological solutions. *Electrochim. Acta* **2004**, *49*, 2167–2178. [CrossRef]

126. Suñer, S.; Tipper, J.L.; Emami, N. Biological effects of wear particles generated in total joint replacements: trends and future prospects. *Tribol. Mater. Surfaces Interfaces* **2012**, *6*, 39–52. [CrossRef]

127. *StanDIN 50320 Wear; Terms, Systematic Analysis of Wear Processes, Classification of Wear Phenomenadards*; Beuth-Verlag: Berlin, Germany, 1979.

128. Chevalier, J.; Taddei, P.; Gremillard, L.; Deville, S.; Fantozzi, G.; Bartolomé, J.F.; Pecharroman, C.; Moya, J.S.; Diaz, L.A.; Torrecillas, R.; Affatato, S. Reliability assessment in advanced nanocomposite materials for orthopaedic applications. *J. Mech. Behav. Biomed. Mater.* **2011**, *4*, 303–314. [CrossRef]

129. Garino, J.P. Ceramic component fracture: trends and recommendations with modern components based on improved reporting methods. In *Bioceramics and Alternative Bearings in Joint Arthroplasty: Proceedings*; D'Antonio, J.A., Dietrich, M., Eds.; Steinkopff: Heidelberg, Germany, 2005; p. 218, ISBN 3798515409.

130. Weisse, B.; Affolter, C.; Stutz, A.; Terrasi, G.P.; Köbel, S.; Weber, W. Influence of contaminants in the stem—ball interface on the static fracture load of ceramic hip joint ball heads. *Proc. Inst. Mech. Eng. Part H J. Eng. Med.* **2008**, *222*, 829–835. [CrossRef]

131. Rehmer, A.; Bishop, N.E.; Morlock, M.M. Influence of assembly procedure and material combination on the strength of the taper connection at the head-neck junction of modular hip endoprostheses. *Clin. Biomech. (Bristol, Avon)* **2012**, *27*, 77–83. [CrossRef]

132. Nevelos, J.E.; Ingham, E.; Doyle, C.; Nevelos, A.B.; Fisher, J. Wear of HIPed and non-HIPed alumina-alumina hip joints under standard and severe simulator testing conditions. *Biomaterials* **2001**, *22*, 2191–2197. [CrossRef]

133. Dorlot, J.-M.; Christel, P.; Meunier, A. Wear analysis of retrieved alumina heads and sockets of hip prostheses. *J. Biomed. Mater. Res.* **1989**, *23*, 299–310. [CrossRef]

134. Mittelmeier, H.; Heisel, J. Sixteen-years' Experience With Ceramic Hip Prostheses. *Clin. Orthop. Relat. Res.* **1992**, *282*, 64–72. [CrossRef]

135. Brandt, J.-M.; Vecherya, A.; Guenther, L.E.; Koval, S.F.; Petrak, M.J.; Bohm, E.R.; Wyss, U.P. Wear testing of crosslinked polyethylene: Wear rate variability and microbial contamination. *J. Mech. Behav. Biomed. Mater.* **2014**, *34*, 208–216. [CrossRef]

136. Grupp, T.M.; Holderied, M.; Mulliez, M.A.; Streller, R.; Jäger, M.; Blömer, W.; Utzschneider, S. Biotribology of a vitamin E-stabilized polyethylene for hip arthroplasty—Influence of artificial ageing and third-body particles on wear. *Acta Biomater.* **2014**, *10*, 3068–3078. [CrossRef]

137. Moro, T.; Takatori, Y.; Kyomoto, M.; Ishihara, K.; Kawaguchi, H.; Hashimoto, M.; Tanaka, T.; Oshima, H.; Tanaka, S. Wear resistance of the biocompatible phospholipid polymer-grafted highly cross-linked polyethylene liner against larger femoral head. *J. Orthop. Res.* **2015**, *33*, 1103–1110. [CrossRef]

138. Zietz, C.; Fabry, C.; Baum, F.; Bader, R.; Kluess, D. The Divergence of Wear Propagation and Stress at Steep Acetabular Cup Positions Using Ceramic Heads and Sequentially Cross-Linked Polyethylene Liners. *J. Arthroplast.* **2015**, *30*, 1458–1463. [CrossRef]

139. Gremillard, L.; Martin, L.; Zych, L.; Crosnier, E.; Chevalier, J.; Charbouillot, A.; Sainsot, P.; Espinouse, J.; Aurelle, J.-L. Combining ageing and wear to assess the durability of zirconia-based ceramic heads for total hip arthroplasty. *Acta Biomater.* **2013**, *9*, 7545–7555. [CrossRef]

140. Reinders, J.; Sonntag, R.; Heisel, C.; Reiner, T.; Vot, L.; Kretzer, J.P. Wear performance of ceramic-on-metal hip bearings. *PLoS ONE* **2013**, *8*, e73252. [CrossRef]

141. Williams, S.; Al-Hajjar, M.; Isaac, G.H.; Fisher, J. Comparison of ceramic-on-metal and metal-on-metal hip prostheses under adverse conditions. *J. Biomed. Mater. Res. Part B Appl. Biomater.* **2013**, *101B*, 770–775. [CrossRef]

142. Halma, J.J.; Señaris, J.; Delfosse, D.; Lerf, R.; Oberbach, T.; van Gaalen, S.M.; de Gast, A. Edge loading does not increase wear rates of ceramic-on-ceramic and metal-on-polyethylene articulations. *J. Biomed. Mater. Res. Part B Appl. Biomater.* **2014**, *102*, 1627–1638. [CrossRef]

143. Al-Hajjar, M.; Jennings, L.M.; Begand, S.; Oberbach, T.; Delfosse, D.; Fisher, J. Wear of novel ceramic-on-ceramic bearings under adverse and clinically relevant hip simulator conditions. *J. Biomed. Mater. Res. Part B Appl. Biomater.* **2013**, *101*, 1456–1462. [CrossRef]

144. Al-Hajjar, M.; Carbone, S.; Jennings, L.M.; Begand, S.; Oberbach, T.; Delfosse, D.; Fisher, J. Wear of composite ceramics in mixed-material combinations in total hip replacement under adverse edge loading conditions. *J. Biomed. Mater. Res. Part B Appl. Biomater.* **2017**, *105*, 1361–1368. [CrossRef]

145. Al-Hajjar, M.; Fisher, J.; Tipper, J.L.; Williams, S.; Jennings, L.M. Wear of 36-mm BIOLOX ® delta ceramic-on-ceramic bearing in total hip replacements under edge loading conditions. *Proc. Inst. Mech. Eng. Part H J. Eng. Med.* **2013**, *227*, 535–542. [CrossRef]

146. Chan, F.W.; Bobyn, J.D.; Medley, J.B.; Krygier, J.J.; Tanzer, M. The Otto Aufranc Award. Wear and lubrication of metal-on-metal hip implants. *Clin. Orthop. Relat. Res.* **1999**, 10–24. [CrossRef]

147. RIPO Annual Report 2016. 2018. Available online: https://www.riotinto.com/documents/RT_2016_Annual_report.pdf (accessed on 1 February 2019).

148. Mayor, S. Registry data show increase in joint replacement surgery. *BMJ* **2017**, *358*, 1. [CrossRef]

149. Kurtz, S.; Ong, K.L.; Schmier, J.; Mowat, F.; Saleh, K.; Dybvik, E.; Kärrholm, J.; Garellick, G.; Havelin, L.I.; Furnes, O.; Malchau, H.; Lau, E. Future clinical and economic impact of revision total hip and knee arthroplasty. *J. Bone Joint Surg. Am.* **2007**, *89*, 144–151. [CrossRef]

150. Kurtz, S.; Ong, K.; Lau, E.; Mowat, F.; Halpern, M. Projections of primary and revision hip and knee arthroplasty in the United States from 2005 to 2030. *J. Bone Jt. Surg Am* **2007**, *89*, 780–785. [CrossRef]

151. Rajeshshyam, R.; Chockalingam, K.; Gayathri, V.; Prakash, T. Reduction of metallosis in hip implant using thin film coating. In *AIP Conference Proceedings*; AIP Publishing LLC: Melville, NY, USA, 2018; Volume 1943, p. 020090.

152. Bijukumar, D.R.; Segu, A.; Souza, J.C.M.; Li, X.; Barba, M.; Mercuri, L.G.; Jacobs, J.J.; Mathew, M.T. Systemic and local toxicity of metal debris released from hip prostheses: A review of experimental approaches. *Nanomed. Nanotechnol. Biol. Med.* **2018**, *14*, 951–963. [CrossRef]

153. Neuwirth, A.L.; Ashley, B.S.; Hardaker, W.M.; Sheth, N.P. Metal-on-Metal Hip Implants: Progress and Problems. In *Biomedical Applications of Metals*; Springer International Publishing: Cham, Switzerland, 2018; pp. 73–93.

154. Abdel Jaber, S.; Affatato, S. *An overview of In Vitro Mechanical and Structural Characterization of Hip Prosthesis Components*; Springer: Berlin, Germany, 2017; ISBN 9783319680255.

155. Lee, J.-M. The Current Concepts of Total Hip Arthroplasty. *Hip Pelvis* **2016**, *28*, 191–200. [CrossRef]

Correction

Correction: Mierzejewska, Z.A. Effect of Laser Energy Density, Internal Porosity and Heat Treatment on Mechanical Behavior of Biomedical Ti6Al4V Alloy Obtained with DMLS Technology. *Materials* 2019, 12, 2331

Żaneta Anna Mierzejewska

Faculty of Mechanical Engineering, Bialystok University of Technology, Wiejska 45c Street, 15-351 Białystok, Poland; a.mierzejewska@pb.edu.pl

Received: 2 September 2019; Accepted: 6 September 2019; Published: 10 September 2019

 check for updates

The authors wish to make the following correction to this paper [1]: The author name "Mierzejewska Żaneta Anna" should be "Żaneta Anna Mierzejewska".

We apologize for any inconvenience caused to the readers.

Conflicts of Interest: The authors declare no conflicts of interest.

Reference

1. Mierzejewska, Ż.A. Effect of Laser Energy Density, Internal Porosity and Heat Treatment on Mechanical Behavior of Biomedical Ti6Al4V Alloy Obtained with DMLS Technology. *Materials* **2019**, *12*, 2331. [CrossRef] [PubMed]

MDPI

St. Alban-Anlage 66

4052 Basel

Switzerland

Tel. +41 61 683 77 34

Fax +41 61 302 89 18

www.mdpi.com

Materials Editorial Office

E-mail: materials@mdpi.com

www.mdpi.com/journal/materials

Lightning Source UK Ltd.
Milton Keynes UK
UKHW050522070720
366110UK00007B/156